T0212711

SpringerBriefs in Applied Sciences and Technology

PoliMI SpringerBriefs

More information about this series at http://www.springer.com/series/11159
http://www.polimi.it

Roham Afghani Khoraskani

Advanced Connection Systems for Architectural Glazing

POLITECNICO
DI MILANO

Springer

Roham Afghani Khoraskani
Politecnico di Milano
Milan
Italy

ISSN 2191-530X   ISSN 2191-5318 (electronic)
SpringerBriefs in Applied Sciences and Technology
ISSN 2282-2577   ISSN 2282-2585 (electronic)
PoliMI SpringerBriefs
ISBN 978-3-319-12996-9   ISBN 978-3-319-12997-6 (eBook)
DOI 10.1007/978-3-319-12997-6

Library of Congress Control Number: 2014957475

Springer Cham Heidelberg New York Dordrecht London

Printed on acid-free paper

Springer International Publishing AG Switzerland is part of Springer Science+Business Media
(www.springer.com)

Preface

Architectural glazing is a widely popular envelope system that was created in the early days of what we now refer to as Modern Architecture. The adoption of glazed systems was basically driven by the transparency and natural illumination that these systems can provide. The highly efficient energy behavior of double-skin façades is one of the major reasons for a growing increase among architects in applying glass skins over their buildings. The aesthetical features of glass also made this material particularly attractive for use in the envelopes of grand and highly invested-in buildings.

Post-earthquake surveys have shown that, although a building designed according to the most contemporary seismic design codes will protect the structure of the building during an earthquake, these provisions are hardly sufficient for avoiding damage to the nonstructural elements of the building. Among the nonstructural elements of a building, those found in glazed envelope systems are among the most vulnerable to damage during an earthquake. This is mainly due to the high rigidity and stiffness of these systems in the in-plane direction, which results in attracting forces, combined with their relative fragility and delicacy with respect to structural members of the building. It has also been observed that the deflections and displacements that occur in the structure of a building during severe loading conditions, such as earthquakes, are likely to be the main cause of damage to a glazed envelope system. In this regard, the seismic behavior of common types of architectural glazing systems has been investigated in this research, and causes of damage to each system have been properly identified. Furthermore, depending on its geometrical and structural characteristics, the ultimate horizontal load capacity of a curtain wall system has been defined based on stability of the glass components. Particular attention has been given to point fixing curtain wall systems where glass panes play a significant role in their structural behavior.

Among different strategies available to minimize the damage to glazed components, the main focus has been to investigate the advantages of incorporating advanced connection devices between the structure of the building and the building envelope system. Different types of connection devices that can be utilized for this purpose are introduced. Advantages and disadvantages of every connection device have been highlighted with regard to both maintaining the integrity of the glazed envelope system, and at the same time protecting it against damaging forces and displacements that occur in the event of a seismic action.

Among different advanced connectors introduced in the literature, the friction damping connections are selected to provide a controllable level of isolation between the envelope system and the structure of the building. This selection is based on the simplicity of their mechanisms and their ability to confine the transferred forces and moments to limited values.

A novel friction connection device that incorporates the friction mechanism between spherical surfaces is introduced, having the advantage of adaptability in almost all glazed envelope systems with complex geometries and high aesthetical demand of its composing elements. And finally, simplified analytical approaches as well as numerical simulations are presented as a basis for tuning the friction connecting devices in glazed systems, which is based on the mechanical strength of the glass panels, connected with friction connectors and their behavior during earthquakes.

Contents

Chapter 1
Introduction

Building envelope systems—especially the façade system—over high-rise build-ings and structures consume approximately over 20 % or more of the total con-struction budget and are considered to be an economically significant attribute of the building. Architectural glass exterior systems, used as the entire building skin or part of its envelope, are considered to be one of the most influential building systems contributing to the proper function of the building. With the exception of a few guidelines in building design codes, there is currently a lack of design approaches provided for designers and engineers in appropriate selection of glazing details to effectively mitigate earthquake damage.

The existing design guidelines for architectural glazing are limited to customary glazed systems in terms of geometry and technology. In many cases they limit the freedom of designers from realizing their desired forms. In order to protect archi-tectural glazing against seismic actions, the concept of offering mechanical com-patibility between the structure of the building and its envelope is examined in this study, in contrast with the common practice of offering clearance between the elements of the envelope system.

We consider that the main cause of damage to glass elements during an earth-quake is the in-plane deformations within the glazed system, generally caused by the deflections and displacements in the structure of the building, Fig. 1.1. Using advanced connection devices, it is possible to avoid these displacements being transferred to the envelope, while still managing to keep the structure of the building as its main support system.

Hence, a set of advanced and energy dissipating connection mechanisms that can be incorporated into building envelope systems are introduced in Chap. 4 and necessary adjustments to make them suitable for architectural glazing systems have been suggested. But prior to that, in order to have a more thorough understanding of the behavior of architectural glazing systems during an earthquake, problems related to seismic behavior of glass and curtain wall systems are presented and investigated in Chap. 3 of this monograph.

© The Author(s) 2015
R. Afghani Khoraskani, *Advanced Connection Systems for Architectural Glazing*,
PoliMI SpringerBriefs, DOI 10.1007/978-3-319-12997-6_1

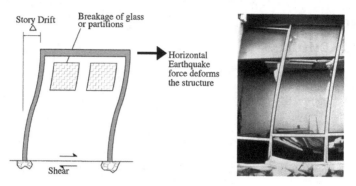

Fig. 1.1 Seismic lateral drifts causing damage to envelope systems, broken glass and bent window mullions in flexible building which experienced large inter-story drift in the 1994 Northridge earthquake (FEMA 310, 1998)

Friction damping devices, often used to reduce the seismic response of structures, are proposed in this research to control the forces applied by the structure of the envelope system. In Chap. 5 a novel friction device is introduced where, in typical sliding friction, connectors are fabricated as rotational friction connectors that have a far greater freedom of movement and ability to be used in more diverse glazed envelope systems as well as providing a higher aesthetical value which is a necessity for transparent systems. In this chapter the evolution of the connection device and the equations and formulas that govern its behavior are investigated.

In Chap. 6, based on the results obtained in Chap. 3 regarding the behavior of glass panels subjected to seismic loading and the equations governing the behavior of the rotational friction connector in Chap. 5, instructions have been provided on tuning and adjusting the behavior of the connection device so as to provide the necessary level of isolation within the envelope system while maintaining the mechanical integrity of the system as much as possible. By properly adjusting the friction connecting devices it is possible to maintain a desired level of isolation between the envelope and the building structure, and to provide mechanical compatibility between the two systems.

During the course of the research which led to preparation of this book, the major objective has been to propose connection systems that result in compatible mechanical behavior between the main structure and the envelope of the building during severe loading conditions. Applicability and adaptability to different glazing systems are considered to be important factors for the proposed systems. The factors below are considered to be major aspects of applicability and adaptability for the connection devices:

- Manufacturing costs
- Simplicity of production and installation
- Ability to be connected to different elements in various positions and orientations
- Having proper size, geometry and aesthetical characteristics

The other objective of the research has been to provide reliable instructions for designers for tuning the friction connections to perform within the expected behavior. This has been achieved by an analytical study of glass panels subjected to the forces applied by friction damping devices. The results of analytical studies of glass panels were later evaluated by numerical simulations.

In this research the main topic of attention is given to safety and avoiding mechanical failures in the glass panels of architectural glazed systems. However, aside from guaranteeing the safety of the glazed system, serviceability and functionality of the envelope would need to be further investigated in terms of air tightness, water tightness and its protection as an acoustic barrier. This will be assured by avoiding damage to the structural elements of the curtain wall system (mullions, transoms, holding brackets etc.) and to the sealant systems (elastic gaskets, silicon patches etc.). In the final chapter of this book a short introduction to experimental testing has been provided both to control and investigate the behavior of the connection device and to check the serviceability of the envelope system that incorporates this connection device—or similar devices—when subjected to severe lateral loading conditions. And finally the conclusions of the research and some notes and suggestions for further developing the topic are presented.

Chapter 2
Architectural Glazing

Abstract It would be difficult to find a material that matches the popularity of glass among engineers and architects. The great importance that glass has attained compared to other materials is associated with its ability to transmit light and to provide a transparent environment. Using this transparent environment and capturing the warmth and brightness of the sun inside a building was a major problem up to the beginning of the twentieth century. In the first half of the twentieth century, due to introduction of new materials to the building industry, such as steel and concrete, as well as more complex structural solutions, integration of glass into the building envelope was made increasingly possible and changed its position from small openings in the façade to considerably large surfaces covering most of the building envelope. In this chapter a brief history of utilization of glass into the building exterior systems is presented. Also different architectural glazing systems that are currently being used are introduced and described.

2.1 From Wall to Skin

Incorporation of glazed elements into the building exterior is parallel to dematerialization of the building façade, which is also coupled with relieving the façade from its load-bearing function and introducing the frame elements which transferred the load to the building foundation. Ultimately, dematerialization of the façades confined their role to being just a skin boundary around the building, separating the inside and outside environments, rather than load-bearing structural members, the boundary that we now refer to as the building envelope (UNI 2005) (Fig. 2.1).

Although before the middle of the twentieth century there have been numerous examples of glass construction, mostly for coverings of railway stations, greenhouses, passages and early modernism of the 20s and 30s that resulted in large private glass houses, the real breakthrough in glass architecture came after world

© The Author(s) 2015
R. Afghani Khoraskani, *Advanced Connection Systems for Architectural Glazing*,
PoliMI SpringerBriefs, DOI 10.1007/978-3-319-12997-6_2

Fig. 2.1 Dematerialization of the façade: **a** 'Auditorium Building' Sullivan and Adler, 1889, **b** Stary Browar shopping center, Poznań, Poland

war II when economic, technological and aesthetic factors all together forced the rapid spread of utilizing glass as a prominent building material. It was at this time that technical advances in glass manufacturing, along with the sensation of modern characteristics embodied in it, made the glazed envelope the symbol of modern architecture. Towering glazed office buildings that were used as the headquarters of giant multinational companies represented growth, confidence and development within those companies, as they still do. Even within city scale a silhouette of high-rise glazed skyscrapers, sharply reflecting the sun during the day and illuminating light as a sign of livelihood during night became signatures of wealth and prosperity (Fig. 2.2).

The leading examples of modern glass architecture were first realized in the United States which, while Europe was dealing with problems of a post war environment, was enjoying the luxury of a reasonable economic growth suitable for investment. More importantly, at the same time, during the years of the Third Reich, it became home to many of the avant-garde immigrants, such as Mies van der Rohe and Walter Gropius (Institut Internationale Architektur-Dokumentation 2007).

Fig. 2.2 New York skyline

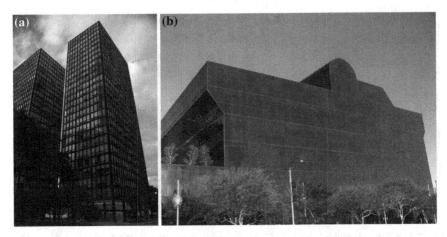

Fig. 2.3 a '860–880 Lake Shore Drive' apartments, Mies van der Rohe; **b** 'Pacific design center' by Cesar Pelli, 1975

Mies van der Rohe, as head of the Illinois Institute of Technology in Chicago and given the task of designing the new university campus, had the chance to reinterpret the idea of a "curtain wall" which later became a distinguished aspect of the high-rise buildings he envisioned. One of the first implementations of glazed curtain wall systems was the façade of his buildings constructed during the 80s along the Lake Shore Drive in Chicago (Fig. 2.3).

Later advances in load-bearing silicon (structural silicon glazing) and other fixing techniques made possible the cladding of rooftops and more complex shapes with glass, allowing a smooth firm skin, eliminating the panel frames and maximizing the glazing of the envelope. The Pacific Design Center designed by Cesar Pelli in 1975 was one of the first structures benefiting from an all glazed envelope.

Other leading examples of modern glass architecture following the Second World War that were realized in the US are: the skyscrapers 'Lever Building', 1951–1952, by the architectural office SOM, the 'Seagram Building', 1954–1958 by Mies van der Rohe, both in New York, and the 'Hancock Tower' in Boston, 1967–1976 by I.M. Pei and Partners in collaboration with H.N. Cobb Fig. 2.4.

While single layer glass panels are the source of considerable heat loss during winter, and direct solar irradiation infiltrating the glazed envelopes may impose considerable energy demand on building services, in the early days of modern architecture, the disadvantageous energy loss in winter and overheating in the summer were resisted passively with tinted glass and by actively using (mainly) energy intensive mechanical air conditioning. After the energy crisis of the 70s double glazing against energy loss and reflective glazing against overheating were increasingly employed. In the meantime the glass industry was able to put double glazing with excellent thermal values on the market, thereby extensively reducing the significance of heat loss. However, undesirable heat gains due to solar radiation continued to pose problems.

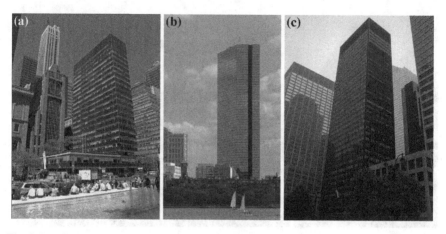

Fig. 2.4 **a** 'Lever Building', 1951–1952, by the architectural office SOM; **b** 'Hancock Tower' in Boston, 1967–1976 by I.M. Pei and Partners; **c** 'Seagram Building', 1954–1958 by Mies van der Rohe

Double-skin facades proved to be a possible solution to this problem. These glazed envelope systems are characterized by the addition of a single glazed skin in front of the double glazed building façade, where a shading device can be located in the cavity between the two façades to control the heat gain due to solar radiation and controlled ventilation between the two layers can be efficiently utilized for the indoor environment.

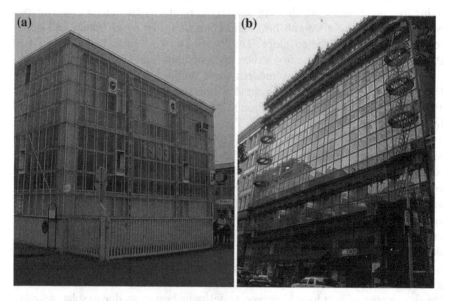

Fig. 2.5 **a** Production Hall Steiff, Giengen on the Brenz, 1903, Architect R. Steiff; **b** 'Hallidie Building', San Francisco, 1915–1917

In the early 90s high quality façade constructions, in particular double-skin building envelopes, were developed and often promoted as the definitive solution to energy loss. Forerunners of double-skin constructions are the projects by Le Corbusier for the not awarded competition design for the People's Palace in Geneva, 1927, and for the building of the Centro-Soyus in Moscow, 1929. The impetus for this solution probably came from traditional double windows which Le Corbusier knew from his country, Switzerland. Double-skin façades had however already been built prior to this but had been recognized as works of civil engineering and not works of architecture. The production hall of the Steiff Company, Giengen on the Brenz, 1903 by R. Steiff, and the 'Hallidie Building', San Francisco, 1915–1917 by W. Polk, are early examples of double-skin glass façades Fig. 2.5.

Double façades are still a very interesting point of discussion in building envelopes in terms of materials, ventilation, shading, translucency and compatibility with the building structure.

2.2 Curtain Walls

A curtain wall is defined as an exterior wall on a building, which does not support the roof or floor loads but is connected to the structural frame, and is an element of the larger building envelope system. While Curtain Walling encompasses systems that use various material claddings such as metal panels and stone, one of the most popular configurations is a metal frame assembly glazed with architectural glass. They usually consist of vertical and horizontal structural members, known as mullions and transoms respectively, connected together and anchored to the supporting structure of the building and in-filled with the cladding panels, to form a lightweight, space enclosing continuous skin, which provides, by itself or in conjunction with the building construction, all the normal functions of an external wall, but does not take on any of the load-bearing characteristics of the building structure (UNI 2005). Glass Curtain wall systems have become a common building component as mass load-bearing wall systems slowly transitioned and were replaced with cavity wall systems during the twentieth century and lightweight wall system options were needed. Curtain walls have many various functions, some of which are harder to achieve effectively than others. These wall systems have requirements which include structural load transfer and resistance, water infiltration protection, air infiltration control, condensation prevention, energy management, sound attenuation, safety, maintainability, constructability, durability, aesthetics, and economic viability (Curtis 1987).

Description of curtain walling types:
The classification of types of curtain walling varies but mainly the following classification is considered which is based on manufacturing, installation and mechanical features of the curtain wall system:

Stick
Unitized
Panelized
Spandrel panel ribbon glazing
Structural sealant glazing
Point-fixed Structural glazing.

2.3 Stick System Curtain Walling

Stick curtain wall system is one of the most primitive and at the same time simple curtain walling systems. In this system the mullions and transoms are long elements mainly from extruded aluminum or cold rolled steel with coating for protection against oxidation and other environmental attacks. The main characteristic of this system is that the components are cut and machined in the factory and the assembly will be completely performed onsite. Figure 2.6 shows the arrangement of the stick curtain wall system. The procedure of onsite assembly is done by first erecting the vertical mullions and connecting them to the structure at the floor slab followed by placement of the horizontal transoms as demonstrated in Fig. 2.7. The distances between the mullions and transoms are defined based on the dimensions of the curtain wall panel. The infill panels will be finally added in between the frames mainly composed of glass panels, however other materials such as metallic or Photovoltaic panels may be used as infill panels. The panels, either working as openings or fixed panels, need to be properly insulated. This will be guaranteed using elastic gaskets that are held in place and under pressure using pressure plates.

Stick curtain walling is a highly adaptable system that can be used for high-rise glass towers as well as single story shop fronts and, although the assembly speed is rather low compared with preassembled systems, it is suitable for irregular shapes

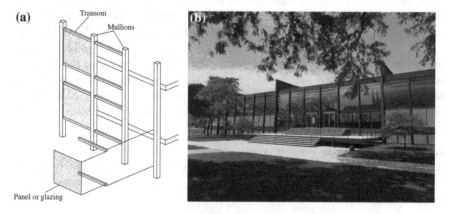

Fig. 2.6 a Stick system curtain walling; **b** IIT Crown hall Mies van der Rohe, 1940

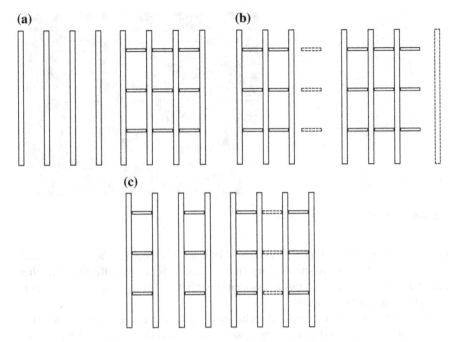

Fig. 2.7 Different construction techniques for stick. **a** Mullions installed then transoms. **b** Mullions and transoms installed sequentially. **c** Ladder frames installed then intermediate transoms

since it can move well with the building frame. The main disadvantage of stick curtain wall systems is associated to its performance and quality control. This is because the assembly is performed onsite and the performance of the systems remains highly dependent on the skills of the assembly and installation team.

2.4 Unitized Curtain Walling

A unitized curtain wall system is currently the most commonly practiced curtain wall system for high-quality glazed envelopes. The comprising modules of unitized system are preassembled glazing panels that are manufactured in controlled factory conditions. These units are composed of a metallic frame—mainly aluminum or steel—that surrounds one or multiple glass layers as well as other materials. The units are usually a story high and connected to the building at every floor slab with prepositioned brackets or a previously installed substructure. The framing system is also positioned to be fastened to the surrounding units with special joints with pressured elastic gaskets to provide an adequately insulated envelope system. Figure 2.8 schematically demonstrates the unitized curtain wall system. Although the framing system in unitized curtain wall is much more complex and the direct

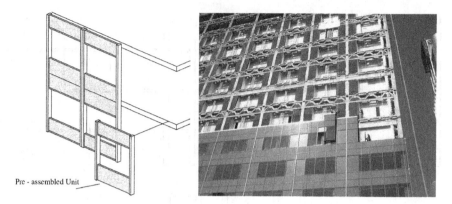

Pre - assembled Unit

Fig. 2.8 Unitized curtain wall systems

cost of the system is higher, through a broader perspective, faster and easier installation and fewer onsite activity makes this system cost effective. Another advantage of unitized systems is associated with their comparatively better performance and quality control.

The installation team of unitized curtain wall systems are usually provided by the manufacturer or selected with its agreement, and instructions need to be provided for proper installation of the system, how to change the damaged units and maintenance of the system. It is also common for these systems to be delivered with some sort of guarantee, depending on the manufacturer.

2.5 Panelized Curtain Walling

Panelized curtain wall systems are quite similar to unitized system with the difference that the modules of panelized curtain wall systems are usually very much larger than unitized systems and they are mostly used when application of the panels directly on the building structure is possible. These modules are again prefabricated panels but with dimensions usually equal to story height in vertical direction and bay span in width, Fig. 2.9. Fixing the panels close to the columns reduces problems due to deflection of the slab at mid-span, which affects stick and unitized systems.

Application of materials other than glass such as metallic plates, composite materials, terra cotta or even precast concrete panels is more common in panelized systems with respect to other curtain wall systems, the panels may even include a substructure framework that is used to support different cladding material like stone and masonry.

Structural steel panelized walls are known as 'truss walls' in North America. Aluminum or galvanized steel skins are generally fixed to the frame with insulation

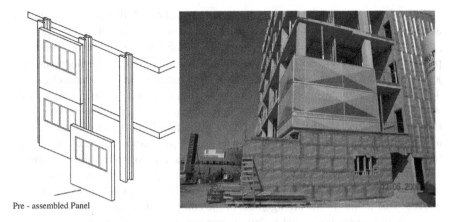

Pre - assembled Panel

Fig. 2.9 Panelized curtain wall systems

in the cavity. The wall construction is then completed by a plasterboard lining and external cladding.

As mentioned earlier, the differences between unitized and panelized systems are mainly associated with dimensions and weight, which is why some choose not to differentiate between the two systems in categorizing curtain wall systems. However, due to fundamental changes that these differences impose on the manufacturing process, connection details, assembly and installation, here they are considered as a separate class of curtain wall systems.

2.6 Spandrel Panel Ribbon Glazing

In spandrel ribbons glazing, long continuous glazed panels are fixed between spandrel panels that are usually fixed to the building floor slab, Fig. 2.10. The main purpose of using spandrel panel ribbon glazing is usually aesthetical and when having a horizontal strip appearance at every story level is desired by the architects.

Spandrel panel

Fig. 2.10 Spandrel ribbon curtain wall systems

The primary support for this system is the spandrel panels that are situated below and above the glazed surface and are fixed to the building floor slab. The spandrel panels may be of various materials such as prefabricated metallic or composite panels or precast concrete units. These units will be fixed to the floor slab via joining brackets and the glazed panels will be fixed between the spandrel panels. The glazed parts of spandrel ribbon curtain wall systems may be assembled onsite with horizontal transoms fixed to spandrel panels that later accommodate glass panes. If necessary for better fitting of the glass panel, vertical mullions may be added in between horizontal transoms. The glazed parts may also be of pre-glazed factory assembled units that will be fixed on bottom and top to the spandrel panels and on the sides to one another. Inserting various types of shading devices, rain screen panels etc. over the glazed surfaces are easily achievable in spandrel ribbon glazing. Due to the special arrangement of the components in this system of glazing it is easy to provide horizontal isolation between the units in different story levels using sliding joints. This can be very helpful for seismic behavior of the curtain wall system as it will be later shown in Chap. 3.

2.7 Structural Sealant Glazing

In structural sealant glazing systems, instead of using mechanical elements such as pressure plates and gaskets, the infill glass panels are fixed to a primary frame system using structural silicon sealants during the factory assembly. Since in structural sealant curtain walls, there is no need for external metallic elements to put the glazing panels in place and under pressure, it is possible to cover almost the total external surface of the envelope with glazing panels. This system of curtain walling is attractive for architects who look for smooth external glazed surfaces. The attachment of these units together and to the main structure can be performed in different ways similar to stick, unitized and spandrel ribbon glazing. The units can be completely preassembled in the factory and similar to unitized systems installed directly on the building structure and fastened together or a substructure of mullions and transoms can be assembled on site and later partly fabricated units composed of glazing panels and a border frame can be bolted on or fastened to the substructure which is similar to a stick curtain wall system, Fig. 2.11.

Structural sealant curtain walls have been used in the United States since the 80s, where in the first proposed systems the glazing panels were directly attached to the fames on site, but due to low performance, inability of quality control and problems concerning maintenance and replacement of damaged panels, this type of assembly is no longer acceptable and the structural silicon sealants need to be performed in the factory and under controlled conditions.

It is customary to assist the structural sealant patches with some mechanical supports, mainly in the form of resting brackets on the lower frame and withholding the weight of the infill panels. This would reduce the size of the sealant bed and

Fig. 2.11 **a** Structural sealant glazing; **b** 'Quay West' building, Manchester by The Ratcliff Partnership Ltd

sometimes make it possible to use structural sealant patches only on two sides and leave the rest to non-structural sealants.

Structural sealant glazing a suitable system for externally complex facet envelopes is commonly referred to as crystal cladding, where the exterior surface is required to be free from protrusions and top over framing. However during the night, and especially when interior lighting is present, the framing system will be visible from outside and the continuously uniform glazed surface will no longer be available. This curtain wall system is usually used on prestigious buildings and may be produced in standard predefined dimensions or tailor cut systems for buildings with irregular and special envelope panels, and as mentioned earlier the framing system may be similar to both stick and unitized systems and any of the previous types of curtain walling and ribbon glazing could incorporate structural silicon glazed elements.

2.8 Point-Fixed Structural Glazing—Bolted Assembly and Patched Assembly

Structural glazing curtain wall systems are envelope systems that provide maximum transparency. The first and foremost feature of these systems is that the support systems have become minimized to achieve maximum external glazing and internal transparency. In both systems, sheets of toughened/laminated glass are assembled with special brackets to a secondary support substructure creating a highly transparent envelope system with a uniform and continuous external surface. These systems are usually incorporated in envelopes which require a minimum amount of framing for a given glass area and as façade and coverings of large internal spaces such as mall entrances, atriums and continuous external skins that are largely separated from the building floor slab and its main structure.

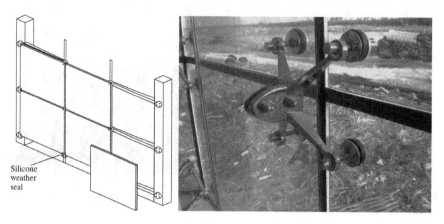

Silicone
weather
seal

Fig. 2.12 Bolted structural glazing

The supporting structure can be in various configurations such as space structures, mullions and transoms, orthogonal substructure net, cable supported structures etc. and all these systems follow a main goal: that is to maximize
transparency. Recently, in close collaboration with the glass industry, achievements
have been made to even use glass pillars as the supporting structure, Fig. 2.12. In
such cases the glass pillars are in the form of long glass plates that act as a support
for forces that are applied normal to the surface of the envelope system as well as its
weight. The height of the glass pillars is equal to the height of the envelope system,
while its thickness and width are designed according to the applied loads predicted
and structural design principles. The envelope panels will be connected at certain
points to the supporting structure using connection brackets. Usually every bracket
supports a number of adjacent panels—typically four—and the gaps between the
panels are weather sealed on-site using wet-applied sealant.

There are two main types of point-fixed structural glazing systems: Bolted
assembly and Patched assembly. In both systems the glazing panels are fixed to the
substructure at the corners of the panels—in case of large panels attachments are
also applied in the middle of the panel borders—, the difference between the two
concerns the fixture of the glazing panels to the attachment brackets. In patched
assembly the corners of the glass panels are held in place by steel brackets applied
at both sides of the glazing panel and held together by applying a pressure. For
better safety usually the lower brackets have underlying flanges for mechanically
withholding the weight of the panels. The brackets are usually in the form of
rectangular patch plates but adopting more intricate shapes for aesthetical reasons is
becoming more common among the manufacturers of point-fixed glazing systems.
After the installation of the glazing panels the space between the adjacent panels
will be weatherproofed on-site using wet sealants or with a combination of elastic
gaskets and wet sealants in case the space between the panels exceeds a few
centimeters, Fig. 2.13.

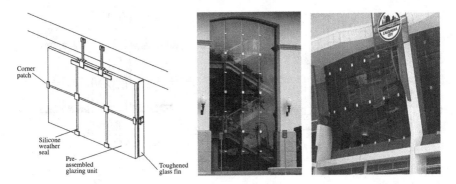

Fig. 2.13 Patch structural glazing

All four borders of the glass panels are sealed with either weather-proof or structural sealants depending on the load transfer system intended by the designers. Although as mentioned earlier mechanical flanges will sometimes be added to assist withholding the weight of the envelope system, the principle behind the design of the fittings in patch plate point glazed envelopes is the friction developed between the Plated/Gasket/Glass interfaces that are produced by the pressure applied over these surfaces by the patch plates. Adhesives may also be applied to enhance the friction between the surfaces.

In contrast to patched assembly systems which mainly rely on friction between certain elements of the fixing system and glass pane, the bolted assembly point-fixed glazed envelopes are connected to their supporting structure completely through mechanical fastening. This would provide a broader margin of safety for using this system in various orientations and non-vertical positions. In bolted assembly systems—commonly referred to as spider systems—the glass panels are provided with holes at their corners and the connection bracket is fastened to the glass through these holes by special bolts. There are two main categories for connection bolts: countersunk fixing and clamp fixing. In clamped fixing the top flange of the bolt is applied over the external surface of the glass, while in countersunk fixing the external end of the bolt is embedded inside the glass and is attached to the fixing hole in a conical shape bringing the end of the bolt in the exact same surface of the glass, Fig. 2.14. In both cases extensive consideration is made to avoid any direct contact between the glass and metallic parts of the bolt which would result in formation and spreading of cracks within the glass pane. Using special bolts that can accommodate more than one layer of glass, it is possible to use double or triple glazed insulated units as the envelope panels.

The entire weight of the glazing panels is transferred to the connection bracket through the bolts. A certain level of flexibility is necessary in the connection bolts, especially with respect to rotation, to avoid stressing the glass after the installation and in the more contemporary designs of the fixings where the bolts remain unconfined in rotary degrees of freedom using ball-type joints. The main design considerations for fixing bolts are described below:

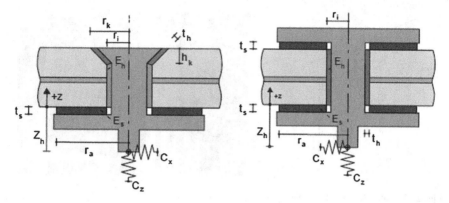

Fig. 2.14 Countersunk fixing (*left*) and clamped fixing (*right*)

Avoiding contact between the metallic components and the glass
Reducing stress concentration in the glass pane near the holes
Providing necessary flexibility to avoid pre-stressing the system
Guaranteeing a uniform distribution of stress at the contacting surfaces around connection points

As formerly mentioned the fixing bolts transfer the forces from the envelope panels to the connection brackets. The connection bracket (spider bracket) consists of usually four arms attached together and fastened to the supporting structure (the number of arms is associated with the number of glass panels connected at every point and the discretization scheme of the envelope). Since every glass panel is independently attached to the support structure there are no limitations regarding height and width for using spider glazing systems, especially when the spaces between the glazing panels is not filled with mechanically resistant sealants and unnecessary exchange of forces that might be the result of thermal expansion or deformations in the substructure is considered improbable.

References

Building Envelope Design Guide. (2009). Homepage of National Institute of Building Sciences, [Online]. Available http://www.wbdg.org/design/envelope.php.
Curtis, W. (1987). *Modern architecture since 1900* (2nd ed.). Englewood Cliffs: Prentice-Hall Inc.
Dutton, H. (1999). Structural glass architecture. In *Proceedings of Glass Processing Days*.
Essiz, O. (2001). Glass facades on steel structures. In *Proceedings of Glass Processing Days*.
Institut Internationale Architektur-Dokumentation. (2007). *Glass construction manual* (2nd Rev. and expanded ed.). Basel, London: Birkhäuser, Springer distributor.
Patterson, M. R. (2008). *Structural glass facades: A unique building technology*. Building Science. Los Angele: University of Southern California.
Pietroforte, R. (1995). Cladding systems: Technological change and design arrangements. *Journal of Architectural Engineering, 1*(3), 100–107.

Schittich, C. (2001). Glass architecture in the second half of the 20th century. In *Glass Construction Manual.*

Schwartz, T. A. (2001). Glass and metal curtain-wall fundamentals. *APT Bulletin, 32*(1), 37–45 (Curtain Walls).

UNI. (2005). *Curtain walling—Product standard.* Italy: UNI—Ente Nazionale Italiano di Unificazione.

The Centre for Window and Cladding Technology [Online]. Available: http://www.cwct.co.uk/home.htm.

Technische Universiteit Delft. Faculteit Bouwkunde. (2008). Challenging glass, Faculty of Architecture, Delft University of Technology, May 2008. In F. Bos, C. Louter & F. Veer (Eds.). *Conference on Architectural and Structural Applications of Glass.* Amsterdam: Delft University Press.

Aughuet, A. A. (1976). Curtain wall structure, no. 3994107.

Vyzantiadou, M. A., & Avdelas, A. V. (2004). Point fixed glazing systems: Technological and morphological aspects. *Journal of Constructional Steel Research, 60*(8), 1227–1240.

Chapter 3
Building Envelope and Mechanical Compatibility

Abstract In this chapter, mechanical compatibility between the building envelope system and the building structure is discussed. Different levels of integration between these two systems are introduced and the problems arising from improper compatibility between them especially in architectural glazing systems are being discussed. The current provision that exist in building design codes for obtaining a compatible behavior and previous studies on achieving mechanical compatibility are presented.

3.1 Building Systems Integration

There are four categories defined in the literature as building systems (Schwartz 2001). These four systems are:

- Structure
- Envelope
- Interior parts
- Mechanical systems and services

There are also other classifications in other references but almost all of them share the ones mentioned above. Each one of these categories is a subject of research and design in different academic and technical fields. Yet another subject of great interest is the relationship between these systems and the influence of one of them on the others and their level of integration. In order to better understand the interactions between a structural system and an envelope it is necessary to define integration levels between these two types of systems. In this study five levels of integration between building systems are considered and introduced as below:

Remote
The case in which the two systems are completely physically separated from one another and are coordinated only in terms of functionality. Examples of this type of

© The Author(s) 2015
R. Afghani Khoraskani, *Advanced Connection Systems for Architectural Glazing*,
PoliMI SpringerBriefs, DOI 10.1007/978-3-319-12997-6_3

integration are noise barrier walls protecting buildings in highly noise polluted areas—such as in airports and highways—having their own supporting structure.

Touching

In this case the envelope of the building has its own supporting structure for carrying its weight but rests on the main structure of the building, relying on it mostly for lateral loads like wind and earthquake. Examples of this type of integration are those with double façades having their own structure and are connected to the main structure of a building with horizontal elements to transfer the lateral loads.

Connected

In this case one system is mechanically fastened and dependent on the other. Examples of this type are those glazed and non-glazed claddings that do not have a support system of their own and are connected to the main structure of the building with brackets and adjusting frames.

Meshed

In this case the two systems are completely attached to each other, nearly occupy the same space, and with no supporting structure for the envelope system. It could be said that the two systems are glued together. Examples of this type of integration are buildings that have stones, ceramics and bricks as their exterior envelope which is glued to the main structure of the building.

Unified

In this case the systems share the same physical components, there are no longer two separate systems, and there is just one physical system that acts both as the building structure and its envelope. Massive concrete structures which benefit from a concrete exterior that is also the main structure of the building are examples of this type.

It is obvious that in the first and the last cases, the mechanical compatibility of a structure with an envelope is meaningless. But in the three middle cases it is very important for the two systems to be well adjusted in two cases of displacement tolerances and deflections (Table 3.1).

Usually, problems with displacement tolerances occur during and before production and construction phases. The displacement tolerances that are accepted during construction of a structural system often tend to centimeters, while considering the great deal of importance given to the exterior envelope of the building in terms of visual appearance, water and air permeability etc., the intensity of tolerances in the envelope is of the order of millimeters. This issue is often tackled using connection details having brackets with slot-holes in three main directions that permit the installation to adjust these tolerances. Other than slot-hole brackets there are other connection details, such as sliding fixings and long rod anchors, that allow installers to fine-tune the placement of the envelope in the right positions.

Dealing with the problems of compatibility between the structure and the envelope for displacements and deformations that occur after installation is much

Table 3.1 Building systems integration diagram

Integrating level	Remote	Touching	Connected	Meshed	Unified
Physical connectivity	No connectivity	Relying on	Fastened with connections	Glued	United[a]

more complicated than adjusting the installation tolerances, mainly because they usually occur after the final fixing of the connections between the two systems and cause stresses and forces in the envelope elements. These deflections that appear in the structure of the building and then are transferred to the envelope are of two types: static and dynamic. The static ones are often the result of added live loads to the structure, such as furnishings or sometimes due to gradual sinking of the building foundation. The dynamic ones are drifts in the structure due to dynamic loads like earthquake and wind forces.

Another problem in adjusting the two systems for deflection displacements is that usually these displacements are not very easy to define. Although structural design codes have various guidelines and restrictions for structural designers from the point of view of deflection, trying to minimize these deflections and preventing them from increasing above certain values, almost all of these restrictions are applied to keep the behavior of the structure within the theoretical designing assumptions and to keep the structure in an operable state. They are far behind the limitations required to maintain the properties of the envelope and other non-structural members. Of course it is reasonable that in extreme situations like earthquakes the center of attention becomes the safety of the structure and its functioning, but in costly and prestigious projects with costs of the exterior of the structure exceeding thousands of Euros per square meter, there must be some considerations to protect these large investments.

3.2 Architectural Glass Provisions in Seismic Codes

The ASCE 7-02 code (Rush, American Institute of Architects 1986) in Section 9.6.2.10 instructs the designers of glazed curtain-walls, store-fronts and glazed partitions to provide enough clearance between the glass panel edges and the frames of such systems to avoid damage to the glass. The following Sections (9.6.2.4.2–9.6.2.10.2) are the verbatim seismic design provisions for architectural glass included in ASCE 7-02. Actual section, equation, table, and reference numbers are cited below for accuracy, and for ease when referring to the ASCE 7-02 document:

1. **9.6.2.4.2 Glass**. Glass in glazed curtain walls and storefronts shall be designed and installed in accordance with section 9.6.2.10 (see below).
2. **9.6.2.8.2 Glass**. Glass in glazed partitions shall be designed and installed in accordance with Section 9.6.2.10 (see below).

9.6.2.10 Glass in Glazed Curtain Walls, Glazed Storefronts, and Glazed Partitions.

9.6.2.10.1 General. Glass in glazed curtain walls, glazed storefronts and glazed partitions shall meet the relative displacement requirement of Eq. (9.6.2.10.1-1)

$$\Delta_{fallout} \geq 1.25\, ID_P, \qquad\qquad (9.6.2.10.1\text{-}1)$$

or 13 mm, whichever is greater,
where $\Delta_{fallout}$ is the relative seismic displacement (drift) causing glass fallout from the curtain wall, storefront or partition; D_p is the relative seismic displacement that the component must be designed to accommodate (Eq. 9.6.1.4-1); D_p shall be applied over the height of the glass component under consideration; and I is the occupancy importance factor (Table 9.1.4).

Exceptions

1. Glass with sufficient clearances from its frame such that physical contact between the glass and frame will not occur at the design drift, as demonstrated by Eqs. (9.6.2.10.1-2-a, 9.6.2.10.1-2-b), shall be exempted from the provisions of Eq. (9.6.2.10.1-1)

$$D_{clear} \geq 1.25\, ID_P, \qquad\qquad (9.6.2.10.1\text{-}2\text{-}a)$$

$$D_{clear} = 2c_1\left(1 + \frac{h_p c_2}{b_p c_1}\right), \qquad\qquad (9.6.2.10.1\text{-}2\text{-}b)$$

   where h_p is the height of the rectangular glass; b_p is the width of the rectangular glass, c_1 the clearance (gap) between the vertical glass edges and the frame; and c_2 the clearance (gap) between the horizontal glass edges and the frame.
2. Fully tempered monolithic glass in Seismic Use Groups I and II located no more than 3 m above a walking surface shall be exempted from the provisions of Eq. (9.6.2.10.1-1).
3. Annealed or heat-strengthened laminated glass in single thickness with interlayer no less than 0.76 mm that is captured mechanically in a wall system glazing pocket, and whose perimeter is secured to the frame by a wet glazed, gun-able curing elastomeric sealant perimeter bead of 13 mm minimum glass contact width, or other approved anchorage system, shall be exempted from the provisions of Eq. (9.6.2.10.1-1).

Discussion of seismic design provisions
In essence, Eq. (9.6.2.10.1-1) requires that the resistance to glass fallout of an individual glass panel be greater than the relative seismic displacement demand the component must accommodate as a result of being attached to the primary structural system of the building. Thus, the resistance to glass fallout is the capacity of a given glazing system. *In the absence of special drift accommodating connections between the main building frame and the curtain wall framing members, this relative seismic displacement demand is governed by the calculated seismic interstory drifts for the specific building being designed for earthquake loading conditions.*

3.3 Energy Dissipation and Mechanical Isolation

The concepts of energy dissipation and mechanical isolation are leading examples of developing innovative concepts to better protect the structure against environmental forces like wind and earthquake or unexpected occurrences like blasts. The design concept of energy dissipation in structural engineering is to absorb or consume a portion of input energy within specific devices in the structural system, thus minimizing the energy that acts upon structural members and protecting them against damage. In the case of mechanical isolation a desired level of isolation is provided between certain elements to avoid them from imposing forces on one another. Both energy dissipation and mechanical isolation are in contrast with the traditional approach that is to provide a combination of strength and ductility to resist the imposed loads. Thus they are being more and more utilized by engineers and designers for protecting buildings from damage during non-permanent environmental and accidental loading conditions.

These design strategies have been thoroughly studied and practiced in the structural design of many buildings, usually to resist seismic actions. Having considered the change of approach in seismic design of buildings from safety criterion to serviceability criterion, more and more attention is now being given to protection of nonstructural elements during an earthquake, among them to the building envelope. There is below a brief study of previous researches in adoption of energy dissipation and system isolation technologies in building cladding and curtain wall systems.

Among the two concepts of energy dissipation and mechanical isolation the former is exclusively practiced in heavy cladding systems while the latter is practiced both in light-weight cladding systems and curtain walls and heavy claddings. The reason for this is that these concepts are historically embedded into the design of buildings mainly by structural designers, and in the case of heavy cladding systems the amount of energy dissipated, which is proportional to the weight of the façade system, is of values that can sometimes significantly affect the seismic behavior of the building structure. So less interest has been given to incorporating energy dissipating devices in light-weight cladding systems although an improvement in their racking performance is expected during cyclic loads. Few researchers have performed experimental tests and researches on the performance of currently used cladding systems.

Rihal (1988a), performed studies of the design and seismic behavior of pre-cast façades/cladding and push-pull connections in low to medium rise buildings with steel frames; his research included experimental testing.

A main and important continuing direction in heavy cladding research is the study of panel frame interaction: a similar and associated subject of research is the use of cladding as an integral part of a lateral bracing system.

Thiel et al. (1986) published the results of a feasibility research on seismic energy absorbing cladding systems. Henry and Roll (1986) performed analytical research of cladding frame interaction. Sack et al. (1989) did studies with experimental testing on cladding/frame interaction. Pall (1989) studied and developed a friction-damped connection for pre-cast concrete cladding. The developed connection has been used in many different projects.

Currently the main researcher in this field is Goodno with his team at Georgia Institute of Technology (Goodno et al. 1989a, b; Pinelli et al. 1993. Goodno et al. 1998). His works include the analytical and experimental studies of different types of "advanced" cladding connectors. The results demonstrated that up to 41 % reduction in peak displacement response could be achieved by configuration of the baseline (as-built) from retrofitting advanced cladding connectors, or 27 % reduction in structural weight (in the longitudinal direction) could be attained for the same baseline response level, Fig. 3.1.

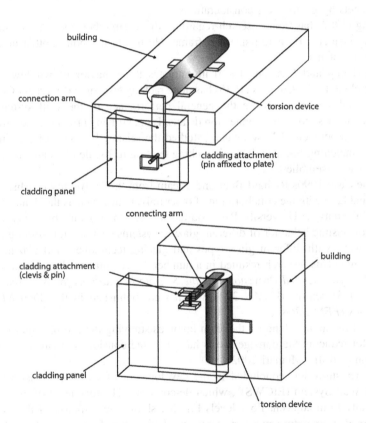

Fig. 3.1 Torsion damping connection *top, horizontal*; *bottom, vertical* (Goodno et al. 1998)

This sophisticated connection device is capable of developing fine energy dissipation behavior through torsion deformation, in a comparable way to the torsion bars in an automobile suspension. The torsion device is made by a circular element mounted inside a concentric tube and attached to a vertical surface of a building structure (in the bottom of Fig. 3.1). The outer tube keeps the torsion element and also fixes it at the lower end, which provides rotational bearing support at the upper end. The connected arm to the torsion element is for converting the inter story drift into rotation (left-right in the figure). The arm is connected with a pin and clevis to the cladding panel (shown in wireframe mode). The upper figure shows a horizontal application.

Among the researches done by different groups on light-weight cladding systems and curtain wall systems, the following are briefly discussed below. As it was earlier indicated, most of these researches focus on the matter of clearance and isolation in curtain wall components, however there are a few that focus on improving the properties of mechanical resistance and avoiding crack formation in glass panels by geometrical considerations.

Wang (1987) has done research regarding the seismic behavior of curtain wall cladding elements on a full-size test frame. He has tested both slotted hole and push-pull systems.

Bouwkamp and Meehan (1960) investigates the behavior of window panels under racking loads. Cupples (1985) has done racking tests on a Robertson-Cupples curtain wall system to assess the general performance of the wall system and estimate the glass-to-frame connection details. Lim and King (1991) at the Building research Association of New Zealand studied the seismic behavior of curtain wall systems, including in-plane dynamic racking tests on full-scale glass and aluminum curtain wall assemblies.

In the early 1990s Richard Behr and a team from the University of Missouri at Rolla, and later with the collaboration of other universities such as the Pennsylvania State University at University Park, started a research program on experimental testing of seismic behavior of different glazing systems such as store-front glazing, curtain walls with different glass types and glazing techniques, and glazing with applied film. This research resulted in a number of recommended revisions for the *NEHRP Recommended Provisions for the Seismic Regulations for New Buildings and Other Structures* (FEMA 302), which were published in the *2000 NEHRP Provisions (FEMA 368)*.

Recent researches done by the team involves studying the use of rounded glass panels for reducing the damage. Tests have resulted considerable gains in accommodation of drift. (Memari 2002).

Another study has concluded in the development of an "Earthquake-Isolated Curtain Wall System (EICWS)", which de-couples each story level of the system structurally from adjacent floor levels Fig. 3.2 shows the response of the isolated curtain wall frames to a number of modes of vibration. Figure 3.3 shows how a

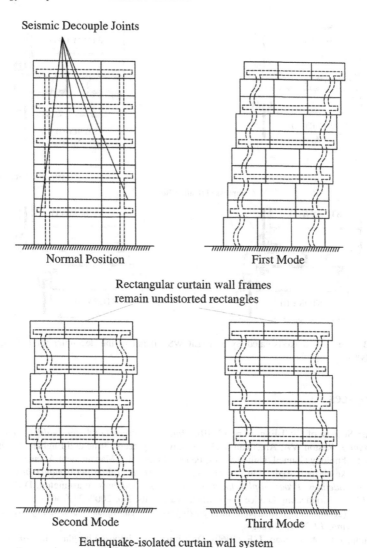

Fig. 3.2 Earthquake-isolated curtain wall system schematic, showing diagrammatic response at first, second and third modes of vibration

seismic "decoupler" joint is able to accommodate relative inter-story movements while still maintaining a building envelope weather seal. In-plane and out-of-plane movements are accompanied by horizontally continuous, flexible, elastomeric gasket loops that act as weather seals between stories.

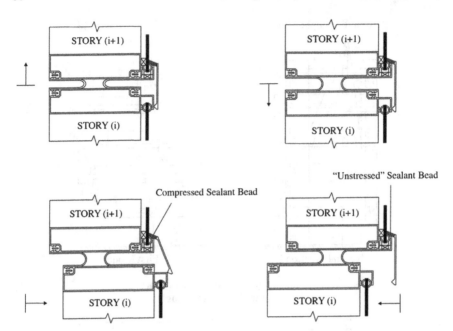

Fig. 3.3 The Earthquake-Isolated System (EICWS), detail of the decoupler joint (Brueggman et al. 2000)

References

American Society of Civil Engineers. (2010). *Minimum Design Loads for Buildings and Other Structures.* Reston, VA: American Society of Civil Engineers: Structural Engineering Institute.

Applied Technology Council, United States. Federal Emergency Management Agency & National Earthquake Hazards Reduction Program. (2006). *Next-generation performance-based seismic design guidelines: Program plan for new and existing buildings.* Washington, D.C.: Fema.

Bai, B. (2009). Connecting Device For Curtain Wall Units. No. 20090249736.

Behr, R. A. (2006). Design of architectural glazing to resist earthquakes. *Journal of Architectural Engineering, 12*(3), 122–128.

Bouwkamp, J. G., & Meehan, J. F. (1960). Drift limitations imposed by glass. In *Proceedings of the Second World Conference on Earthquake Engineering*, Tokyo (pp. 1763–1778).

British Standards Institution. (1996). *Eurocode 8: Design provisions for earthquake resistance of structures. Pt. 1.1, General rules: Seismic actions and general requirements for structures.* London: British Standards Institution.

Brueggeman, J. L., Behr, R. A., Wulfert, H., Memari, A. M., & Kremer, P. A. (2000). Dynamic racking performance of an earthquake-isolated curtain wall system. *Earthquake Spectra, 16*(4), 735–756.

Cupples (1985). Curtain Wall Tests for Cupples Horizon Series of World Wall. Cupples test report No. STL–33, St. Louis, MO.

De Gobbi, A. (2010). *Curtain wall anchor system.* US7681366 ed. US.

Evans, D., & Lopez Ramirez, F.J. (1989). Glass damage in the 19 September 1985 Mexico earthquake.

Goodno, B., Zeevaert-Wolff, A., & Craig, J. I. (1989a). Behavior of heavy cladding components. *Earthquake Spectra, 5*(1), 195–222.

Goodno, B. J., Craig, J. I., & Zeevaert Wolff, A. (1989b). *Behavior of architectural nonstructural components in the Mexico earthquake. Final progress report.*

Goodno, B. J., Craig, J. I., Dogan, T., & Towashiraporn, P. (1998). *Ductile Cladding Connection Systems for Seismic Design.* Gaithersberg, MD: Building and Fire Research Laboratory, NIST.

Heng, H. (2004). *Design of structural glass fitting for seismic condition.* Toowoomba, Australia.

Henry, R. M., & Roll, F. (1986). Cladding-frame interaction. *Journal of Structural Engineering New York, N.Y., 112*(4), 815–834.

Hsu, C. C., Calise, A. J., Sweriduk, G. D., Goodno, B. J., & Craig, J. I. (1994). *Building seismic response attenuation using robust control and architectural cladding. Proceedings of the American Control Conference* (p. 1073).

Lessons Learned from the 1985 Mexico Earthquake 1989.

Lilli, D. (2009). System for fixing panels, slabs, glass walls, etc. to supporting surfaces in the building field and/or in the furniture field. No. 20090199509.

Lim, K., & King, A. B., (1991). The behavior of external glass systems under seismic in-plane racking, *Building Research Association of New Zealand (BRANZ) Study Report,* 39.

Memari, A. M., Kremer, P. A., & Behr, R. A. (2002). Architectural glass panels with rounded corners to mitigate earthquake damage. Dept. of Architectural Engineering, Pennsylvania State University, University Park PA.

Memari, A. M., Behr, R. A., & Kremer, P. A. (2003). Seismic Behavior of Curtain Walls Containing Insulating Glass Units. *Journal of Architectural Engineering, 9*(2), 70–85.

Memari, A. M., Kremer, P. A., & Behr, R. A. (2006). Architectural glass panels with rounded corners to mitigate earthquake damage. *Earthquake Spectra, 22*(1), 129–150.

National Technical Information Service. (1995). Literature Review on Seismic Performance of Building Cladding Systems. Gaithersburg, MD.

Pantelides, C. P., & Behr, R. A. (1994). Dynamic in-plane racking tests of curtain wall glass elements. *Earthquake Engineering and Structural Dynamics, 23*(2), 211–228.

Pantelides, C., Deschenes, J., & Behr, R. (1993). Dynamic in-plane racking tests of curtain wall glass components. *Structural Engineering in Natural Hazards Mitigation,* 664.

Pantelides, C. P., Truman, K. Z., Behr, R. A., & Belarbi, A. (1996). Development of a loading history for seismic testing of architectural glass in a shop-front wall system. *Engineering Structures, 18*(12), 917–935.

Pall, A. S. (1989). Friction-Damped Connections for Precast Concrete Cladding. In *Proceedings, Architectural Precast Concrete Cladding—Its Contribution to Lateral Resistance of Buildings,* PCI, Chicago, IL. pp. 300–309.

Pinelli, J. P., Craig, J. I., Goodno, B. J., & Hsu, C. C. (1993). Passive control of building response using energy dissipating cladding connections. *Earthquake Spectra, 9*(3), 529–546.

Pinelli, J., Craig, J. I., & Goodno, B. J. (1995a). Energy-based seismic design of ductile cladding systems. *Journal of Structural Engineering, 121*(3), 567–578.

Pinelli, J. P., Craig, J. I., & Goodno, B. J. (1995). Energy-based seismic design of ductile cladding systems. *Journal of Structural Engineering—ASCE, 121*(3), 567–578.

Pinelli, J. P., Moor, C., Craig, J. I., & Goodno, B. J. (1996). Testing of energy dissipating cladding connections. *Earthquake Engineering and Structural Dynamics, 25*(2), 129–147.

Rihal, S. (1988a). Seismic Behavior and Design of Precasar Facades/Cladding and Connections in Low/Medium Rise Buildings. Report ASCE R88-1, Arch. Eng. Dept. School of Arch. And Envir'l Design, CA Polytechnic University, San Luis Obispo, CA.

Rush, R. D., & American Institute of Architects. (1986). *The building systems integration handbook.* Chichester, New York: Wiley for the American Institute of Architects.

Sack, R. L., Beers, R. J., & Thomas, D. L., (1989). Seismic behavior of architectural precast cladding. In *Proceedings, Architectural Precast Concrete Cladding-Its Contribution to Lateral Resistance of Buildings,* PCI, Chicago, IL. pp. 141–158.

Schwartz, T. A. (2001). Glass and metal curtain-wall fundamentals, *APT Bulletin,* Curtain Walls, *32*(1), 37–45.

Starr, C. M., & Krauthammer, T. (2005a). Cladding-structure interaction under impact loads. *Journal of Structural Engineering, 131*(8), 1178–1185.

Starr, C. M., & Krauthammer, T. (2005b). Cladding-structure interaction under impact loads. *Journal of Structural Engineering, 131*(8), 1178–1185.

Starr, C. M., & Krauthammer, T. (2005c). Cladding-structure interaction under impact loads. *Journal of Structural Engineering, 131*(8), 1178–1185.

Thiel, C. C., Elsesser, E., Lindsay, J., Bertero, V. V., Filippou, F., & McCann, R. (1986). Seismic energy absorbing cladding system: A feasibility study. In *Proceedings, ATC-17 Workshop and Seminar on Base Isolation and Passive Energy Dissipation*, ATC, San Francisco, CA, pp. 251–260.

Wang, M. L. (1987). Cladding performance on a full-scale test frame. *Earthquake Spectra*, EERI, *3*(1), 119–173.

Wulfert, H. (2003). Earthquake-immune curtain wall system. No. 6598359.

Chapter 4
Seismic Behavior of Glass Curtain Walls

Abstract As mentioned earlier, the basic cause of damage to the nonstructural elements of the building during an earthquake is deflections and displacements within the structural elements. In this chapter the behavior of the glass panels subjected to these displacements are investigated for different types of architectural glazing systems. For every system depending on the fixtures and their boundary conditions, the maximum allowable lateral force that may be exerted on the panel is derived using the theory of plates for the three cases of dry glazed systems, structural sealant systems and point fixed systems. Simplification has been made in order to achieve analytical solutions.

4.1 Seismic Behavior

In order to protect the nonstructural elements of buildings, seismic design codes provide limitations on story drift during the structural design phase (British Standards Institution 1996; International Code Council 2000). The story drift ratio is provided in two main directions of the story plane, and in each direction is defined as the relative displacement between the top and bottom of the story divided by the story height. These limitations are mostly based on psychological comfort of the inhabitants during severe situations (avoiding large swinging in floors) and serviceability of typical construction technologies (avoiding failure in partition walls and mechanical appliances). These limitations are hardly enough for glass façades and further considerations need to be made.

During the structural design phase, in order to investigate the effect of deflections within the structure of the building over the glass facades and envelopes of the building, it is advised that the structural engineers provide a list of these deformations and drifts from their numerical simulations and analysis results for different load cases and combinations. However in case of absence of such data the maximum allowable drift ratios in the seismic code used for the structural design will be used as input data. In this research—in accordance with most seismic design codes—0.02 is

© The Author(s) 2015

R. Afghani Khoraskani, *Advanced Connection Systems for Architectural Glazing*,
PoliMI SpringerBriefs, DOI 10.1007/978-3-319-12997-6_4

considered as maximum allowable drift ratio for the structural design, and a value of 0.01 is considered as a usually witnessed drift ratio in buildings in event of an earthquake.

4.2 Dry Glazed Systems with Edge Clearance

The behavior of glass panels remains in elastic range both during the in-plane and out-of-plane deformations. The main causes of damage to glass panels during an earthquake are the in plane deformations which occur in the curtain wall system. This is due to significant stiffness of the glass panels in that direction. Sucuoğlu and Vallabhan (1997) argues that the in-plane deformation response of a glazed curtain wall having clearance provided between the glass panel edges and the frame can be described in two phases. In the first phase the glass panel undergoes a rigid body motion within its supporting frame, without having any force exerted on it. In the second phase the two opposite corners of the glass panel start to experience a diagonal pressure from the supporting frame.

4.2.1 Deformation Due to Rigid Body Motion

Figure 4.1a schematizes a glass panel and its supporting frame having c as clearance between the panel edge and the frame. It can be observed in Fig. 4.1b, c that the panel can withstand a relative displacement of D_r according to the clearance length and the dimensions of the panel without having exerted any force on the glass panel; the displacements of the glass are only due to rigid body motion.

Equation (4.1) shows the relationship between allowable panel deformation (D_r) and the panel size (h_p and b_p) and clearance (c),

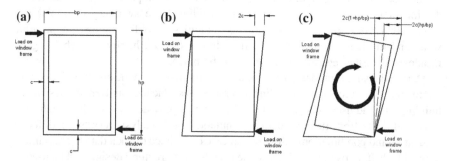

Fig. 4.1 In-plane deformation within a curtain wall system due to rigid body motion; **a** undeformed shape, **b** glass edges in contact with the frame, **c** glass corners under pressure by the frame

Table 4.1 Rigid motion allowable displacement for different glass panel sizes and clearances

Height (m)	Width (m)	Thickness (mm)	c (cm)	D_r (cm)	Drifts
1	1		0.5	2	0.02
			1	4	0.04
1.3	0.85		0.5	2.53	0.019
			1	5.06	0.039
1.5	1.2		0.5	2.25	0.015
			1	4.5	0.03
2	1		0.5	3	0.015
			1	6	0.03
2.6	1.7		0.5	2.53	0.01
			1	5.06	0.019

$$D_r = 2c\left(1 + \frac{h_p}{b_p}\right). \tag{4.1}$$

The above concept is more generalized in ASCE 7-02 (American Society of Civil Engineers 2010) as it was previously described in Sect. 3.2. It can be observed that within this approach the mechanical strength of the glass panels, to withstand load, is not brought into consideration and the design objective has been to avoid contact between the glass edges and the frames. The requirement described by ASCE 7-02 Section 9.6.2.10 is either mentioned or referred to in numerous building design codes. Table 4.1 shows the in-plane displacements that can be endured within a window panel having edge clearances equal to 0.5 and 1 cm for different glass panel sizes; it is assumed that both vertical and horizontal clearances have the same value. The drift values associated to every glass pane is derived by the endured displacements divided by the height of the glass pane. These values are not dependant on the thickness of the glass pane and only relate to its dimensions.

4.2.2 Deformation Due to Pressure on the Glass

What we have previously presented occurs in the first phase of the deformation capacity of a glazed system, which is due to the rigid body motion of the glass inside the frame and provided by the clearance between the glass edges and window frame. There is also another deformation component, in addition to the rigid body component expressed previously, that contributes to the in plane deformation capacity of window glass.

When the glazed system reaches the ultimate deformation that can be achieved with rigid body motion (already expressed by D_{clear}), Fig. 4.1c the two opposite corners of the glass panel coincide with the adjacent corners of the window frame, and the glass plate will start to receive diagonal compression. At this stage, it tends to deflect in a diagonal buckling mode accompanied by a shortening in the loaded diagonal direction and further rotation together with the window frame as shown in Fig. 4.2. A lateral deflection D_r results due to this shortening/rotation mechanism, which can be related to the diagonal shortening $\Delta d = d-d'$ through simple geometric relationships by assuming $D_r \ll d$,

$$\Delta d = \frac{b}{d} D_r. \tag{4.2}$$

The shortening in the glass panel is the result of an out-of-plain buckling deformation in a diagonal direction which is expressed in Fig. 4.3. If we consider that a deformed shape function of the glass panel follows a sinusoid shape, as in the buckling deflection of long columns under longitudinal loads we have:

$$y = a \sin \frac{\pi x}{d}, \tag{4.3}$$

where a is the maximum amplitude at the centre, and Δd the diagonal shortening of the glass panel along the applied forces.

Sucuoğlu and Vallabhan (1997) defined the in-plane deformation in the curtain wall panel caused by diagonal shortening of the glass panel with the assumption that the glass will reach its maximum allowable stress in the middle of the panel by Eq. (4.4)

$$D_r = \frac{1}{b} \left(\frac{\sigma_{all} d^2}{\pi E t} \right)^2. \tag{4.4}$$

Table 4.2 shows the values for D_r for different panel sizes and thicknesses based on Eq. (4.4).

Fig. 4.2 In-plane deformation within a curtain wall system due to glass panel deformation

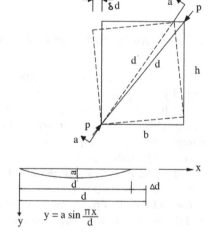

Fig. 4.3 Glass panel deformation along the diagonal of the plate (Section a-a in Fig. 4.1)

Table 4.2 Allowable displacement by mechanical deflection for different glass panel sizes and thicknesses

Height (m)	Width (m)	Thickness (mm)	D_r (cm)	Drifts
1.00	1.00	6	0.574966	0.00575
1.00	1.00	10	0.206988	0.00207
1.30	0.85	6	0.984233	0.007571
1.30	0.85	10	0.354324	0.002726
1.50	1.20	6	1.630999	0.010873
1.50	1.20	8	0.917437	0.006116
1.50	1.20	12	0.40775	0.002718
2.00	1.00	8	2.021365	0.010107
2.00	1.00	16	0.505341	0.002527
2.60	1.70	8	4.429048	0.017035
2.60	1.70	10	2.834591	0.010902
2.60	1.70	16	1.107262	0.004259

The elastic modulus (E) equal to 70 GPa and allowable stress (σ_{all}) equal to 50 GPa has been considered as mechanical properties of glass panels.

4.3 Unitized and Panelized Systems

Unitized and panelized systems are curtain wall systems that are mostly manufactured in factories and then assembled and adjusted onsite. Although due to more complex framing and added components these systems have higher costs, the increase in the construction speed and also better sealing and enclosure performance of these systems, due to higher quality control over production achieved in factories, has made them a more often practiced solution for curtain walling compared with stick framing systems. Being produced in a better controlled environment—a factory—makes the utilization of structural silicon as both the sealant and holding components of these systems trouble-free. Structural silicone is commonly used in these systems to fasten the glass panes to the framing which is then attached to the main structure of the building.

Having the glass pane fully fastened to the framing system by structural silicon makes it impossible to provide necessary clearance between the glass pane and the framing system. On the other hand, having supported all the edges of the glass pane evenly provides a uniform distribution of the loads applied on the glass and prevents the effect of localized stresses.

The seismic provisions considered for these types of curtain walls, which are not always put into practice, are essentially based on the concept of separating them from the building structure in the horizontal in-plane direction. Non-dissipating connections (introduced in Chap. 6) are usually used in these systems for protecting the curtain wall against lateral drifts. In the event of absence of isolating connectors or their improper behavior these systems may experience three types of damage:

1. Frame distortions
2. Structural silicon rapture
3. Glass pane failure

The first two types of damage are related to the serviceability of the curtain wall but the third type relates to safety measures, unless safety glass (laminated or tempered) are used as composing glass panes. The effects of the seismic forces over the structural silicon and the framing system can only be investigated by performing experimental tests on real-scale curtain wall mock-ups, and the results of such tests can highly differ according to the materials used and details of the composing elements. But for glass panes subjected to shear forces around the edges it is possible to use the theory of buckling of plates to achieve the maximum load-bearing capacity of the glass within the system.

4.3.1 Plate Buckling for Glass Panes Subjected to Shear

In structural silicon glazing, drift deformations in the curtain wall system do not cause compression forces on opposite corners of the glass panels as described by Vallabhan for dry glazed systems. Instead, due to the continuity of the connections in the glass panels these deformations act as shear forces on their edges. So in order to investigate the behavior of the glass panels in a structural silicon curtain wall the behavior of plates subjected to shear force along the sides needs to be studied (Fig. 4.4).

Considering the large dimensions of glass panels compared with thickness, the most probable failure mode of a glass panel subjected to shear force along its edges is shear buckling.

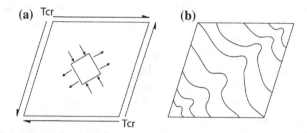

Fig. 4.4 Plate subjected to shear loading; **a** principal stresses in the center of plate, and **b** deformation in the plate

Having the governing differential equation of plates as below:

$$\frac{\partial^4 w}{\partial x^4} + 2\frac{\partial^4 w}{\partial x^2 \partial y^2} + \frac{\partial^4 w}{\partial y^4} = \frac{1}{D}\left(q + N_x\frac{\partial^2 w}{\partial x^2} + 2N_{xy}\frac{\partial^2 w}{\partial x \partial y} + N_y\frac{\partial^2 w}{\partial y^2}\right)$$

$$D = \frac{Et^3}{12(1-v^2)}$$

(4.5)

where: w is the out-of-plane deflection, q is the out-of-plane pressure acting on the plate, t is the thickness of the panel, E and v are respectively the modulus of elasticity and Poisson ratio.

The governing differential equation of plates is used as the basis for determining the critical shear force acting on the glass edges which causes the glass panel to buckle. The only difference is that in the governing differential equation (Eq. 4.5) we only consider the N_{xy} and the rest of the external forces are assumed to be zero. Hence the Eq. (4.5) will be simplified to:

$$\frac{\partial^4 w}{\partial x^4} + 2\frac{\partial^4 w}{\partial x^2 \partial y^2} + \frac{\partial^4 w}{\partial y^4} = \frac{2}{D}\left(N_{xy}\frac{\partial^2 w}{\partial x \partial y}\right).$$

(4.6)

Using the simplest form of the Fourier series, which is compatible with the deformation in the glass pane and its boundary conditions, we can assume w to be in the form of:

$$w = \sum_m \sum_n w_{mn} \sin\frac{m\pi x}{a} \sin\frac{n\pi y}{b}.$$

(4.7)

This satisfies the boundary condition of simply supported plates. Now in order to find the critical values for N_{xy} we need to substitute w from Eq. (4.7) to Eq. (4.6) and solve the resulting differential equation for its obvious solution which gives us the unstable situation of the equation and in our case buckling. The following is the obvious solution to the resulting differential equation:

$$(N_{xy})_{cr} = -\frac{abD}{32}\frac{\sum_m \sum_n a_{mn}^2\left(\frac{m^2\pi^2}{a_2} + \frac{n^2\pi^2}{b^2}\right)^2}{\sum_m \sum_n \sum_p \sum_q a_{mn}a_{pq}\frac{mnpq}{(m^2-p^2)(q^2-n^2)}}.$$

(4.8)

It is necessary to select such a system of constants a_{mn} to make N_{xy} a minimum. This will be done by equating the derivatives of the Eq. (4.8) with respect to a_{mn}. Using the proper notation we can have:

$$(N_{xy})_{cr} = k\frac{\pi^2 D}{b^2} \rightarrow \tau_{cr} = k\frac{\pi^2 D}{b^2 t},$$

(4.9a)

where k is a constant depending on the ratio $a/b = \beta$ and has to be numerically calculated in order to minimize Eqs. (4.9a, 4.9b). Limiting the number of considered mode shapes to 5, the values of k are given in Table 4.3.

The value of k for the case of simply supported plates subjected to shear force, can also be estimated with accuracy of more than 90 % with the below formulas;

$$k = 5.34 + \frac{4}{(h/b)^2}; \quad (h/b) \geq 1$$

$$k = 4 + \frac{5.34}{(h/b)^2}; \quad (h/b) \leq 1$$

(4.9b)

It is now possible to determine the shear capacity of the glass panels and the forces that can be applied on them through the connection devices. Table 4.4 shows the values of the shear capacity for different sizes and thicknesses of glass panels.

As obvious from the table the glass panes, while supported with structural silicon and subjected to shear forces, can demonstrate a considerable amount of strength against buckling and therefore failure. It is much more probable that the damage will be imposed on the silicon patches or the framing system of the unitized and panelized glazing systems before the glass panes start to experience damage. Although the effects of localized stresses and local buckling, which also contribute to the failure mode of the glass panel, especially the region near the connection devices, need to be further investigated.

Table 4.3 Values of factor k in the Eq. (4.11)

a/b	1.0	1.2	1.4	1.5	1.6	1.8	2	2.5	3	4
k	9.4	8.0	7.3	7.1	7.0	6.8	6.6	6.1	5.9	5.7

Table 4.4 Values of critical shear force for different window panel sizes and thicknesses

Height (m)	Width (m)	Thickness (cm)	k	τ_{cr} (kg/cm^2)	N_{xy} (kg/cm)
1.00	1.00	0.60	9.35	204.4063	122.6438
1.30	0.85	0.60	7.060059	91.32809	54.79686
1.50	1.20	0.60	7.91	76.85581	46.11349
1.50	1.20	0.80	7.91	136.6326	109.306
2.00	1.00	0.8	6.35	61.69841	49.35873
2.60	1.70	0.8	7.060059	40.59026	32.47221
2.60	1.70	1	7.060059	63.42229	63.42229

4.4 Point-Fixed Glazing Systems

The idea behind point-fixed structural glazing systems, either bolted assembly or suspended assembly, is a high aesthetical demand for an all-glass envelope from the outside view and maximum transparency from inside. The fact that a continuous and smooth glazed surface is a design objective and requirement for these curtain wall systems eliminates the possibility of providing isolations between the glass panels in horizontal and vertical directions. Also the fact that the glass panes, in these systems, are fully fastened to the supporting structure and will be directly subjected to the damaging loads in the appearance of lateral drifts in the main structure highlights the necessity to separately investigate the glass pane behavior with respect to applied forces and displacements.

4.4.1 Glass Pane Buckling for Structural Glazing

Again considering the comparatively small values of glass thickness compared to its height and width, the theory of plates is used for analyzing the behavior of glass panes in structural glazing and, as well as most structural plate members subjected to in-plane loads, the case of buckling is the most probable form of failure. Thus we need to make sure that the imposed forces do not exceed the critical values for plate buckling.

For such an analysis first we consider that the glass pane is connected at its four corners to the substructure. It is possible to exert the effect of lateral movements and drifts with a pair of transversal loads at the top corners of the plate which results in the reaction forces at the bottom corners of the plate as shown in Fig. 4.5.

In this case the effect of the acting forces on the glass pane is analogous to the case of glass pane subjected to shear, but instead of having the shear equally distributed along the edges of the plate it is applied at two points at the two ends of plate edges (Fig. 4.6).

It should be noted that the total amount of the distributed shear force, which is the shear intensity multiplied by the length of the glass edge and its thickness, is equal to the sum of the point forces applied at the glass corners.

One major difference in this case with the case of plate subjected to distributed shear forces, earlier discussed in Section (3-2), is that in this case the boundary conditions of the plate are free, contrary to the plate subjected to distributed shear forces where the boundary conditions were supposedly considered to be simply supported. This change in the boundary conditions will result in considerable differences in the value of k and it is not possible to adapt the values of plate buckling coefficient from Table 4.3. In this case the value of k is considered to be equal to 0.43, which is an all-time minimum value for buckling coefficient of plates and is used when the boundary conditions of the plate do not play any role in resisting the buckling of the plate. This value will be used for all sizes and thicknesses of glass

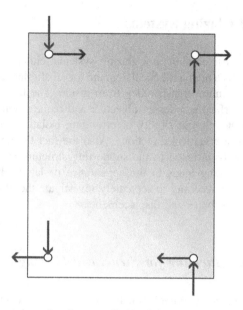

Fig. 4.5 The exerting and reaction forces applied by fixings of the panel caused by lateral drifts

Fig. 4.6 Replacing the point loads on the fixing holes of glass panel with equivalent distributed shear on the edge

panes. The values of this analysis on the buckling of glass are presented in Table 4.5.

In the Table 4.5; k is the buckling coefficient, τ is the imaginary distributed shear, F_h and F_v are the horizontal and vertical components of the maximum allowable force on the glass pane corners and are based on the formulations below:

$$\tau_{cr} = k\frac{\pi^2 D}{b^2 t}$$
$$F_h = \frac{\tau_{cr} \times b \times t}{2} \qquad (4.10)$$
$$F_v = \frac{\tau_{cr} \times b \times t}{2} \times \frac{h}{b}.$$

Table 4.5 Values of critical point loads on the fixings of spider glazing glass panes with different sizes and thicknesses

Height (m)	Width (m)	Thickness (cm)	k	τ_{cr} (kg/cm^2)	F_h (kg)	F_v (kg)
1.00	1.00	0.60	0.43	9.400506	282.0152	282.0152
1.00	1.00	1.00	0.43	26.11252	1,305.626	1,305.626
1.30	0.85	0.60	0.43	5.562429	141.8419	216.9347
1.30	0.85	1.00	0.43	15.45119	656.6757	1,004.328
1.50	1.20	0.60	0.43	4.178002	150.4081	188.0101
1.50	1.20	0.80	0.43	7.42756	356.5229	445.6536
1.50	1.20	1.2	0.43	16.71201	1,203.265	1,504.081
2.00	1.00	0.8	0.43	4.178002	167.1201	334.2402
2.00	1.00	1.6	0.43	16.71201	1,336.961	2,673.922
2.60	1.70	0.8	0.43	2.472191	168.109	257.1078
2.60	1.70	1	0.43	3.862798	328.3378	502.1638
2.60	1.70	1.6	0.43	9.888763	1,344.872	2,056.863

Another approach for analytically demonstrating the effect of the applied forces over structural glazing glass panels is to investigate the case where a slightest rotation occurs in the panel due to lateral drifts in the structure, and in this case the glass pane encounters a diagonal pressure imposed over two of its opposing corners as in Fig. 4.7.

Here it can be shown that the stress state resulting in a plate from such assumption is very much similar to the case of glass pane subjected to distributed shear but with free edge conditions (Fig. 4.8).

Since there is no closed form solution for plates under diagonal pressure, it is assumed that the glass panel in question is part of an imaginary rectangular plate parallel to its diagonal, as is demonstrated in Fig. 4.9. It is also assumed that the force p is constantly distributed along the edges perpendicular to the diagonal under compression with intensity N_p. Having these assumptions it is now possible to calculate the maximum allowable force P causing the compression buckling in a glass panel.

Another approach is then presented here for calculating the critical buckling forces which is the buckling of column plates subjected to pressure. This is the case when a rectangular plate is free over two edges and simply supported on the other two sides where a uniform pressure is applied on the plate, Fig. 4.10.

According to the well-known Euler theory, the critical buckling load for any column becomes:

$$P_{cr} = \frac{\pi^2 \cdot EI}{h^2}.$$

(4.11)

Fig. 4.7 Glass pane
subjected to diagonal loading
from opposite corner fixings

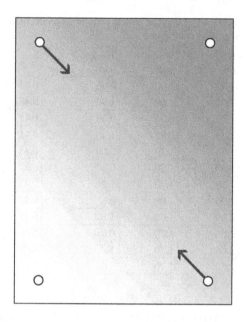

Fig. 4.8 Similarity between
plates subjected to shear
loading and diagonal pressure

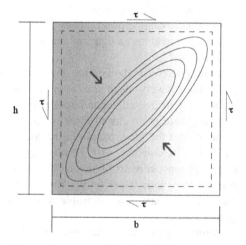

For adjusting this expression for relatively large width in relation to the buckling length of the strut we need to add a quotient $1/(1 - v^2)$, which is due to the free strain deformation in the transverse direction in the center part, in relation to constraints at the loaded edges. And we get:

$$P_{cr} = \frac{\pi^2 \cdot EI}{h^2} \cdot \frac{1}{(1 - v^2)}. \tag{4.12}$$

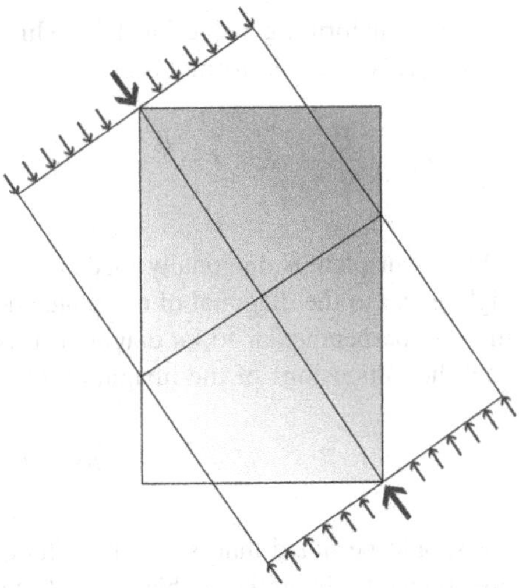

Fig. 4.9 Imaginary plate-
beam subjected to uniform
pressure replacing the initial
diagonally loaded glass plate

Fig. 4.10 Plate-column with
uniform pressure loading and
simply supported

Now transforming the critical buckling load (P_{cr}) to an equivalent buckling stress (σ_{cr}) we get the formulas below:

$$\sigma_{cr} = \frac{P_{cr}}{b \cdot t} \quad \& \quad I = \frac{b \cdot t^3}{12} \quad \rightarrow \quad \sigma_{cr} = \frac{\pi^2 \cdot E}{12 \cdot (1 - v^2) \cdot \left(\frac{h}{t}\right)^2}. \tag{4.13}$$

Since our plate is diagonally loaded, the imaginary plate is now assumed with a height equal to the diagonal of our plate and the width equal to the cross section of our plate perpendicular to its diagonal. Based on the geometry of the initial glass panel the dimensions of the imaginary rectangle are defined below:

$$h' = \sqrt{b^2 + h^2}, \quad b' = \sqrt{b^2 + h^2} \cdot \frac{b}{h}. \tag{4.14}$$

It should be noted that, since the solution to the newly proposed problem is an upper bound to the initial problem, the data calculated can be conservatively used in a design process, however the effect of local stresses and local buckling on the edges of the glass panel shall be separately considered in a numerical modeling (Fig. 4.11).

The results of this analysis are presented in Table 4.6 for different sizes and thicknesses of glass panes and then compared to the results of the first approach.

Fig. 4.11 Imaginary plate-beam subjected to uniform pressure replacing the initial diagonally loaded glass plate

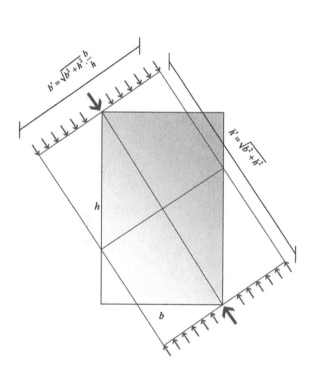

Table 4.6 Values of critical diagonal pressure on the opposing corners of spider glazing glass panes

Height (m)	Width (m)	Thickness (m)	h'	b' (kg/cm^2)	P (kg)	F_h (kg)	F_v (kg)
1.00	1.00	0.60	1.41	1.41	927.51	655.85	655.85
1.00	1.00	1.00	1.41	1.41	4,294.03	3,036.34	3,036.34
1.30	0.85	0.60	1.55	1.02	552.17	302.18	462.15
1.30	0.85	1.00	1.55	1.02	2,556.36	1,398.97	2,139.60
1.50	1.20	0.60	1.92	1.54	546.27	341.25	426.57
1.50	1.20	0.80	1.92	1.54	1,294.87	808.90	1,011.13
1.50	1.20	1.2	1.92	1.54	4,370.19	2,730.04	3,412.55
2.00	1.00	0.8	2.24	1.12	695.24	310.92	621.84
2.00	1.00	1.6	2.24	1.12	5,561.93	2,487.37	4,974.74
2.60	1.70	0.8	3.11	2.03	654.43	358.14	547.74
2.60	1.70	1	3.11	2.03	1,278.18	699.48	1,069.80
2.60	1.70	1.6	3.11	2.03	5,235.43	2,865.08	4,381.89

Knowing that the horizontal component of the force P results from the drifts in the structure and is the origin of damage to the glass, it can be seen that this amount shows good proximity to the sum of the horizontal components of the first approach, over the top and bottom edges of the glass. The same is also true for the vertical components of both approaches.

4.5 Laminated Glass Corrections

All the calculations above have been made under the assumption that glass shows homogeneous characteristics and also behaves with respect to the Bernoulli theory for bending which indicates that, during the course of bending, the sections of the beam, column, plate etc. remain flat plates. This is not entirely true for the case of laminated glass where an interlayer of PVB is present between two layers of glass, Fig. (4.12).

In this case the ability of the middle section of the glass to completely transfer the shear stresses developed parallel to the interlayer will highly drop, and results in a much less bending stiffness in the glass panel. One way to overcome this issue is by describing the load-carrying behavior of the glass according to the sandwich theory (Dweib et al. 2004; Bedon and Amadio 2012). This way the critical buckling pressure for the case of plate-column glass with a two-layer sandwich becomes:

Fig. 4.12 Schematic figure of laminated glass section

$$N_{cr} = \frac{\pi^2(1 + \alpha + \pi^2\alpha\beta)}{1 + \pi^2\beta} \cdot \frac{EI_s}{h^2}$$

$$\alpha = \frac{I_1 + I_2}{I_s}$$

$$\beta = \frac{t_{PVB}}{G_{PVB} \cdot b(z_1 + z_2)^2} \cdot \frac{EI_s}{h^2} \qquad (4.15)$$

$$I_i = \frac{bt_i^3}{12}$$

$$I_s = b(t_1 z_1^2 + t_1 z_2^2).$$

Although being analytically very accurate the problem with the above approach is that it is limited to the problem of plates with uniaxial loading having two free and two simply supported edges. Another way to include the effect of layered glasses in the buckling analysis that was previously described is to come up with an equivalent thickness for laminated glass panels (Fig. 4.13).

Assuming that the PVB interlayer does not provide any shear resistance in between the glass layers of the same thickness, we observe that the overall boundary stiffness of the glass will drop by 75 % compared to the case of having a single layer glass with the same thickness.

$$I_1 = 2\frac{b \cdot t_1^3}{12}, \quad I_2 = \frac{b \cdot t_2^3}{12} = 8\frac{b \cdot t_1^3}{12} \quad \rightarrow \quad I_1 = 0.25I_2. \qquad (4.16)$$

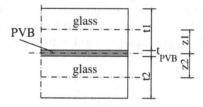

Fig. 4.13 Laminated glass section and proportions

Now if we assume a single layered glass with the same bending stiffness of the double layered glass, the thickness of that single layered glass will be the equivalent thickness of our glass pane:

$$I_1 = \frac{b \cdot t_{eq}^3}{12} = 2\frac{b \cdot t_1^3}{12}$$

$$\rightarrow t_{eq} = \sqrt[3]{2t_1} = 1.26t_1, t_{eq} = \sqrt[3]{0.25}t_2 = 0.63t_2 \tag{4.17}$$

In cases where shear stress is developed in laminated glass parallel with the interlayer, the interlayer can be considered as having some shear resistance. This can be taken into account in evaluating resistance to bending of the laminated glass using a suitable engineering formula in combination with the shear resistance of the interlayer. Herein the following approach, using the concept of 'effective thickness' can be used:

The effective thickness for calculating bending deflection is:

$$h_{ef;w} = \sqrt[3]{(1 - \varpi) \sum_i h_i^3 + \varpi \left(\sum_i h_i\right)^3}. \tag{4.18}$$

And the effective thickness for calculating the stress of glass layer number j is:

$$h_{ef;\sigma;j} = \sqrt{\frac{\left(h_{ef;w}\right)^3}{(h_j + 2\varpi h_{m;j})}} \tag{4.19}$$

where ϖ is a coefficient between 0 and 1, representing no shear transfer (0) and full shear transfer (1),

h_i, h_j are the thicknesses of the glass ply, Fig. (4.14), and

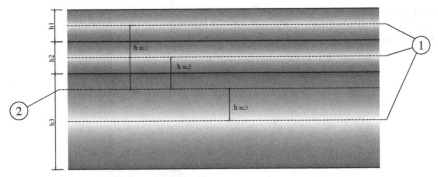

1 Mid-plane of each glass ply

2 Mid-plane of laminated glass

Fig. 4.14 Example of 3 layers laminated glass thickness dimensions

Table 4.7 Effective thicknesses of laminated glass with two plies of the same thickness and $\varpi = 0.25$

Glass thickness	Short duration loads ($\varpi = 0.25$)		Long duration loads ($\varpi = 0.05$)	
	$h_{ef;w}$	$h_{ef;\sigma;j}$	$h_{ef;w}$	$h_{ef;\sigma;j}$
3 + 3	4.55	5.02	3.96	4.44
4 + 4	6.07	6.69	5.28	5.92
5 + 5	7.59	8.37	6.60	7.40
6 + 6	9.11	10.04	7.92	8.88
8 + 8	12.15	13.39	10.56	11.84
10 + 10	15.18	16.73	13.20	14.80

$h_{m;j}$ is the distance of the mid-plane of the glass ply j from the mid-plane of the laminated glass, ignoring the thickness of the interlayer, Fig. (4.14).

The effective thicknesses for calculating stresses and deflection in laminated glass comprising two plies of the same thickness using a value of $\varpi = 0.25$ are given in Table 4.7.

4.5.1 Determination of ϖ

The value of ϖ to be used for a specific interlayer and a particular load case depends on the interlayer stiffness family to which the interlayer belongs for that particular load case. The interlayer stiffness families and the equivalent values of ϖ are given in Table 4.8.

Based on the corrections on the laminated glass thickness, the values of the tables presented in sections above for critical buckling forces of glass panels for the two assumptions of distributed shear buckling and diagonal pressure presented in the Tables 4.9 and 4.10.

Table 4.8 Value of ϖ associated with an interlayer stiffness family

Interlayer stiffness family	Value of ϖ
4	0.7
3	0.5
2	0.25
1	0.1
0	0

Table 4.9 Values of critical point loads on the fixings of spider glazing glass panes with different sizes and thicknesses (laminated glass)

Height (m)	Width (m)	Thickness (cm)	τ_{cr} (kg/cm^2)	F_h (kg)	F_v (kg)
1.00	1.00	0.60	5.57356	128.7492	128.7492
1.00	1.00	1.00	15.48211	596.0612	596.0612
1.30	0.85	0.60	3.297964	64.75553	99.03787
1.30	0.85	1.00	9.161012	299.7941	458.5087
1.50	1.20	0.60	2.477138	68.66626	85.83282
1.50	1.20	0.80	4.4038	162.7645	203.4556
1.50	1.20	1.2	9.908551	549.33	686.6626
2.00	1.00	0.8	2.477138	76.29584	152.5917
2.00	1.00	1.6	9.908551	610.3667	1,220.733
2.60	1.70	0.8	1.465762	76.74729	117.3782
2.60	1.70	1	2.290253	149.8971	229.2543
2.60	1.70	1.6	5.863048	613.9784	939.0257

Table 4.10 Values of critical diagonal pressure on the opposing corners of spider glazing glass panes (laminated glass)

Height (m)	Width (m)	Thickness (cm)	P (kg)	F_h (kg)	F_v (kg)
1.00	1.00	0.60	423.44	299.42	299.42
1.00	1.00	1.00	1,960.37	1,386.19	1,386.19
1.30	0.85	0.60	252.09	137.95	210.99
1.30	0.85	1.00	1,167.06	638.67	976.80
1.50	1.20	0.60	249.39	155.79	194.74
1.50	1.20	0.80	591.15	369.29	461.61
1.50	1.20	1.2	1,995.14	1,246.35	1,557.94
2.00	1.00	0.8	317.40	141.95	283.89
2.00	1.00	1.6	2,539.20	1,135.57	2,271.13
2.60	1.70	0.8	298.77	163.50	250.06
2.60	1.70	1	583.53	319.34	488.40
2.60	1.70	1.6	2,390.15	1,308.01	2,000.48

References

Aiello, S., Campione, G., Minafò, G., & Scibilia, N. (2011). Compressive behaviour of laminated structural glass members. *Engineering Structures, 33*(12), 3402–3408.

Bedon, C., & Amadio, C. (2012). Buckling of flat laminated glass panels under in-plane compression or shear. *Engineering Structures, 36*, 185–197.

Behr, R. A. (2006). Design of architectural glazing to resist earthquakes. *Journal of Architectural Engineering, 12*(3), 122–128.

Bradford, M. A., & Azhari, M. (1995). Buckling of plates with different end conditions using the finite strip method. *Computers & Structures, 56*(1), 75–83.

British Standards Institution (1996). *Eurocode 8: Design provisions for earthquake resistance of structures. Pt. 1.1, General rules: Seismic actions and general requirements for structures.* London: British Standards Institution.

Building Research Association (1989). *The behavior of external glazing systems under seismic in-plane racking.* Building Research Association of New Zealand.

Carré, H., & Daudeville, L. (1999). Load-bearing capacity of tempered structural glass. *Journal of Engineering Mechanics, 125*(8), 914–921.

Chopra, A. K. (2001). *Dynamics of structures: Theory and applications to earthquake engineering* (2nd ed.). Upper Saddle River, NJ, London: Prentice Hall International.

European Standards. (2009). *prEN 13474-3.* Brussels: European Standards.

Glass Processing Days (2005). In *Proceedings of International Conference on Architectural and Automotive Glass* Tamglass, Tampere, Finland. June 17–20 2005.

Hayman, B., Berggreen, C., Lundsgaard-Larsen, C., Delarche, A., Toftegaard, H., Dow, R. S., et al. (2011). Studies of the buckling of composite plates in compression. *Ships and Offshore Structures, 6*(1–2), 81–92.

Heng, H. (2004). *Design of structural glass fitting for seismic condition.* Australia: Toowoomba.

Lee, S., Yoon, S.J., & Back, S.Y. (2006). Buckling of composite thin-walled members. *Key Engineering Materials, 326*, 1733–1736.

Luible, A., & Crisinel, M. (2004). Buckling strength of glass elements in compression. *Structural Engineering International: Journal of the International Association for Bridge and Structural Engineering (IABSE), 14*(2), 120–125.

O'Brien, W. C. (2009). *Development of a closed-form equation and fragility curves for performance-based seismic design of glass curtain wall and storefront systems.* Department of Architectural Engineering, The Pennsylvania State University.

Schlaich, J., Schober, H., & Moschner, T. (2005). Prestressed cable-net facades. *Structural Engineering International: Journal of the International Association for Bridge and Structural Engineering (IABSE), 15*(1), 36–39.

Shi, G., Zuo, Y., Shi, X., Shi, Y., Wang, Y., & Guo, Z. (2010). Influence of damages on static behavior of single-layer cable net supported glass curtain wall: Full-scale model test. *Frontiers of Architecture and Civil Engineering in China, 4*(3), 383–395.

Sucuoğlu, H., & Vallabhan, C. V. G. (1997). Behaviour of window glass panels during earthquakes. *Engineering Structures, 19*(8), 685–694.

Timoshenko, S. (1961). *Theory of elastic stability* (2nd ed.). New York: McGraw-Hill.

Young, W. C. (2002). *Roark's formulas for stress and strain* (7th ed.). New York, London: McGraw-Hill.

Zarghamee, M. S., Schwartz, T. A., & Gladstone, M. (1996). Seismic behavior of structural silicone glazing. *ASTM Special Technical Publication, 1286*, 46–59.

Chapter 5
Advanced Connectors

Abstract The use of advanced connectors in cladding systems has been proposed by many scholars and designers after post-earthquake surveys. Laboratory tests had shown that fixed elements of a cladding system are vulnerable to damage during an earthquake due to deformation accruing in the structure of buildings. The idea of using advanced connectors was to provide isolation between the envelope system and the structure and to dissipate seismic energy. Since light-weight cladding systems do not affect the dynamic behavior of the building, giving very little contribution to it, it is obvious that the energy dissipating approach on a building scale can only be carried out in heavy cladding systems. Goodno et al. (Ductile Cladding Connection Systems for Seismic Design NIST, Gaithersberg, 1998) provide a detailed study of different dissipating connection systems. But since energy dissipating mechanisms can also be used as a means of controlling the forces resulting from displacements, they still have the potential for being used in light cladding systems in order to provide a desirable level of isolation. Due to their simplicity, both in terms of analytical study and practical use and high control over the forces that are transmitted, friction damping connectors are proposed in this research as suitable connecting devices between the glazed envelope and the structure of the building.

5.1 Advanced Connectors

Although advanced connectors that are used in cladding systems have considerable differences based on the systems they are used in, and the behavior which is expected from them, there are some primary functional conditions that they all face and some general characteristics which they all need to have. Cladding connections must be able to transmit the following sets of applied loads:

Vertical loads
Normal loads
Transversal loads

© The Author(s) 2015
R. Afghani Khoraskani, *Advanced Connection Systems for Architectural Glazing*,
PoliMI SpringerBriefs, DOI 10.1007/978-3-319-12997-6_5

The vertical loads, quite obviously, are the ones which act in the vertical direction over the connections and they are usually caused by the weight of the cladding elements, but they may also be the result of thermal expansions within the cladding panels or other occurrences in the cladding system. Transmitting the vertical load is the main function of a connection in cladding systems and therefore is its most important responsibility. The normal loads are the ones that are applied in the direction normal to the surface of the envelope and they are mostly caused by wind and crowd loads and sometimes by the acceleration effect of the movement of the cladding elements during earthquakes. Transmitting the normal forces will keep the cladding systems in place. The transversal loads are the ones that are acting in the plane of the cladding system in the horizontal direction and they are mainly caused by the displacements of inter-story drifts during earthquakes, although they can also be caused by thermal expansion in the envelope surface, or in case of heavy weight cladding by acceleration nature of the seismic forces. The main function of advanced connectors for cladding systems is to isolate the panels against transversal loads by insuring that all the connectors are flexible or energy dissipating in the transversal direction. This will guarantee that although the cladding system remains intact with the structure in normal conditions, isolation can be provided in transversal direction between the two systems to protect the envelope against inter-story drift and if possible dissipate seismic energy in the building during an earthquake.

Applied loads over connection systems are not necessarily confined to the forces acting on the connection systems and in the case of presence of significant rotational constraints, these forces will result in moments applied over connection elements.

Advanced connectors are assumed to function in every aspect the same as conventional cladding connectors, as they will also be able to transmit transversal loads between the two systems as much as the panels can support. However with the proper design of the connectors it should be possible to limit the maximum applied in-plane force that is introduced to the panel in order to protect the panel and its attachments. However different the connectors may be, they are generally composed of three main components:

  Anchor to the envelope
  Connection body
  Anchor to the structure

Based on the type of the cladding used as the building envelope, the structural system of the building (and in some cases the secondary structure of the envelope system), the architectural requirements and the function of the connection, the components of the connection device have considerable design variations. Usually the energy dissipating mechanism within a connection device happens in the connection body.

Different mechanisms of advanced connectors:
Providing the desired properties of energy dissipation and transversal isolation, that are the main purposes of advanced connectors, can be achieved by adapting different mechanical approaches. Maintaining the structural integrity of the connection

device during repeated cycles of transverse deflections, having a well-defined force transmitting value and a bulky stable hysteresis loop are some major characteristics that should be taken into account for every mechanism and material selected. It is only by satisfying the conditions above that an advanced connector can perform in a predictable behavior which guarantees the protection of the envelope system.

In general there are four types of advanced mechanical approaches used in advanced connectors to satisfy the conditions introduced above:

5.1.1 Yield Damping Connectors

The inelastic deformation of the metallic substances is one of the most effective mechanisms available for seismic energy dissipation in structures. In these connection devices the post yield ductility of the metallic members provides the required energy dissipation in one of the three mechanisms of torsional bending, flexural beam bending and U-strip flexural bending; Fig. 5.1 demonstrates the concept of the three introduced mechanisms.

These systems (especially the flexural bending mechanism) have already been used as seismic energy damping mechanisms in the bracings of steel structure.

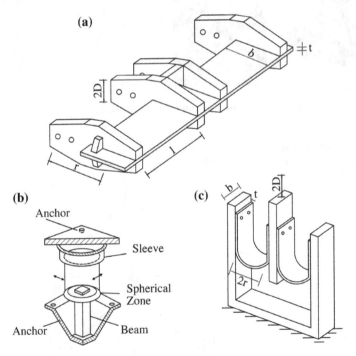

Fig. 5.1 Mechanisms of yield damping connectors; **a** flexural beam bending, **b** torsional bending, **c** U-strip flexural bending

During the ensuing years, considerable progress has been made in the development of metallic dampers. For example, many new designs have been proposed, including the X-shaped and triangular plate dampers displayed in Fig. 5.2. Alternative materials, such as lead and shape-memory alloys, have been evaluated. Numerous experimental investigations have been conducted to determine performance characteristics of individual devices and laboratory test structures. As a result of this ongoing research program, several commercial products have been developed and implemented in both new and retrofit construction projects. In particular, a number of existing structures in New Zealand, Mexico, Japan, Italy, and the United States now include metallic dampers as a means for obtaining improved seismic resistance.

With some adjustments in geometry, shape and especially size, the connections with yield-damping concept that are utilized in the passive dynamic control of the structures can be modified to be used as advanced connection devices for claddings. Figure 5.3 shows the modified yield-damping connections to be used as advanced connection devices in building cladding.

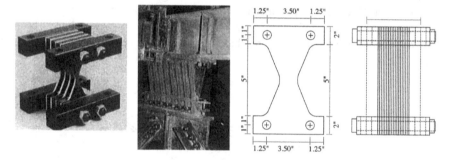

Fig. 5.2 Metallic dampers; X-shaped plate and triangular plate damper

Fig. 5.3 Triangular plate damper within structural frame, brace-damper assembly

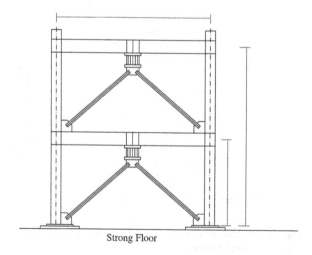

Strong Floor

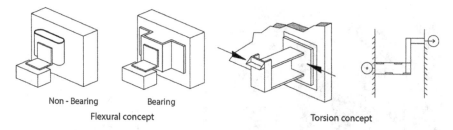

Non - Bearing   Bearing

Flexural concept   Torsion concept

Fig. 5.4 Yield damping connectors for cladding

The employment of yield-damping connections in the building cladding is limited to heavy cladding systems like concrete claddings or terracotta cladding panes, where the yielding of the metallic plates can be easily achieved (Fig. 5.4). In light cladding systems in order to adapt the yield-damping connectors the thickness and the dimensions of the metallic components must become so small that they will endanger the structural integrity of the connection device.

5.1.2 Visco-elastic Shear Based Connectors

The visco-elastic dampers were first utilized in vibration control of aircraft back in the 1950s in order to reduce the members fatigue caused by vibrations in the system. Their first applications in the building sector dates back to 1969 when 10,000 visco-elastic dampers where used in late twin towers of the world trade center to help resist the wind loads followed by a number of similar adaptations to tall buildings against wind loads. Implementation of these systems for seismic purposes has a more recent origin since due to their frequency dependent behavior it requires a much more effective use of these mechanisms (Fig. 5.5).

Figure 5.6 shows a typical version of VE connections in which a visco-elastic layer in bounded between steel plates, shear deformation accompanied by energy dissipation occurs when structural drifts induce a relative deformation between the adjacent steel plates. Copolymers or glassy substances that dissipate energy when subjected to shear deformations are usually materials used in visco-elastic layers.

Studying the basic behavior and the mechanical governing equations for these systems shows that the shear storage modulus and the loss modulus of visco-elastic materials are generally dependent on excitation frequency, ambient temperature, shear strain and the material temperature. Having that said, it can be concluded that employment of these connection devices in building envelopes for seismic considerations will result in loss of adequate predictability due to the varying ambient temperature—which is in direct contact with the building envelope—and the uncertain behavior during an earthquake. However based on the visco-elastic shear connections that are typically utilized for base isolation in bridge supports, some connection devices have also been introduced for the cladding systems.

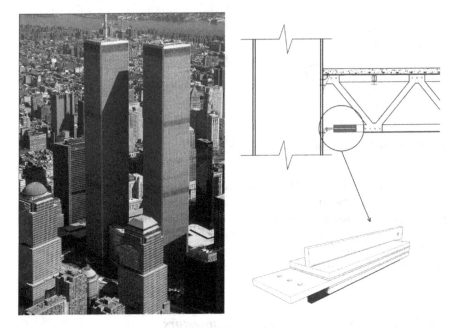

Fig. 5.5 Visco-elastic shear connectors in WTC twin towers

Fig. 5.6 Schematic
demonstration of a typical VE
damping connector

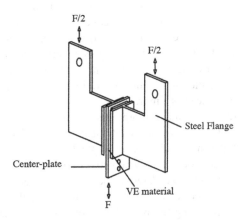

The design objective of these connections has been to develop a load-bearing connector that provides panel alignment and maintain high levels of ductility in the lateral direction while supporting the panel weight. For this, it is required to have high levels of stiffness in two directions and a lower level with hysteresis behavior in the transverse direction. To that end elastonomer neoprene bearing pads laminated with steel plates are used in these connections. They are very rigid in the direction of bearing but capable of sustaining large displacements perpendicular to

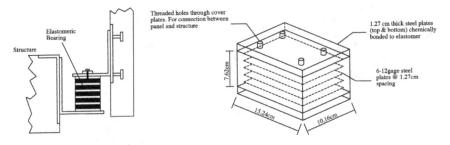

Fig. 5.7 Visco elastic shear damping connectors for cladding systems

it. The steel laminates prevent the elastomeric material from bulging under large compressive forces, but play no role in the shear stiffness of the connection which depends only on the elastomeric material and pad dimensions. Figure 5.7 demonstrates a typical scheme of these connection devices.

5.1.3 Friction Damping Connectors

In the previous connection mechanisms the energy dissipating behavior that had been considered as the basis for the design of the connections occurred in the bulk of material of the composing elements. In the friction connection mechanism the energy dissipating action which occurs between the surfaces of two contacting surfaces is used as the concept for designing the connection device.

Friction is one of the most influencing physical properties of materials and has been employed in many natural and also engineering processes. So it is of no surprise that this phenomenon has been used to solve a variety of engineering problems. Many studies have been performed on investigating the friction between two contacting surfaces, both on microscopic and macroscopic levels, but the coulomb model for friction has been shown to be an adequate explanation of the trend in most of the engineering problems at hand.

The friction mechanism is the basis of a great number of energy dissipating devices in structural engineering. A well-known example of these devices is the slotted bolted connection introduced by FitzGerald et al. shown in Fig. 5.8.

Another major difference between the friction connections and advanced connectors previously discussed is that, unlike the yielding and visco-elastic connections, the friction connector's force displacement diagram does not follow a smooth curve but changes the behavior quite rapidly. Prior to the point that the applied forces on the connection reach a maximum value, the connection device will remain as a rigid connection point, but when the forces reach the maximum value the forces transferred through the connection will remain constant and equal to the maximum design value and the connection will begin to experience relative movements

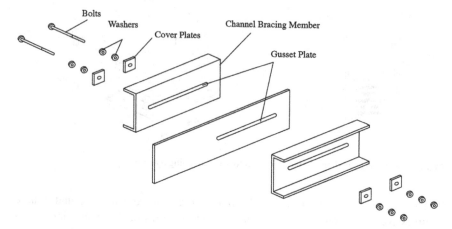

Fig. 5.8 Slotted bolted connection (FitzGerald et al. 1989)

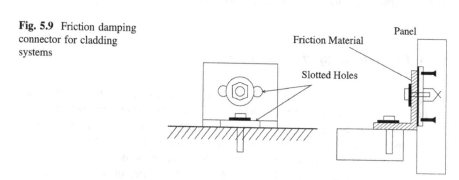

Fig. 5.9 Friction damping connector for cladding systems

between its two ends. Figure 5.9 demonstrates a typical scheme of the friction connection device used in cladding systems. More detailed information on friction connection devices is presented in the next section of this chapter.

5.1.4 Non-dissipating Isolators

Some of the highly practiced connection mechanisms to protect the building envelope during seismic actions are the ones that do not rely on the energy dissipating behavior of the connections and focus the attention on the isolating feature of the connection. Since, as it was already mentioned earlier, light-weight cladding systems do not play a significant role in the overall dynamic response of the building in case of an earthquake, the adoption of non-dissipating isolators has been more limited to these envelope systems. But in some heavy cladding systems considerations are made in order to separate the cladding from the building structure, both for protecting the cladding system during an earthquake and for avoiding unwanted and uncalculated effects of the cladding system over the building structure.

There are two main categories of non-dissipating isolating connection devices in building cladding systems; rocking connections and swaying connections. In the rocking connection systems the connection devices are managed in a way that the cladding panels can endorse a rotating movement in case lateral drift occurs in the main structure, Fig. 5.10. But in the swaying connection systems the cladding connections are managed so that the panel can freely experience translation in a horizontal direction. In both cases usually the lower connection provides resting points for the panel while the upper connections provide resistance only against out-of-plain movements. But this is not necessarily always this way and sometimes the situation is vice versa, and the panels are hanging from their top connections.

In rocking connection systems tie-back connections are typically used for minimum stiffness against lateral movement and high resistance against out-of-plain movements while the panel is resting on or hung from load-bearing supports. In the sway connection systems, slotted-hole connection systems replace the tie-back connections in order to restrain the movements only to transversal directions.

Sway connection systems are highly preferred in glass curtain wall, since rotation in the glass panels can result in distortions in the curtain wall system and equal horizontal displacements at story level is much easier to mechanically deal with in such systems. Figures 5.11 and 5.12 demonstrate an advanced sway connection system for curtain walls (Brueggeman et al. 2000), introduced by Wulfert et al. in which the panels are only connected to one floor level at every story and decoupling

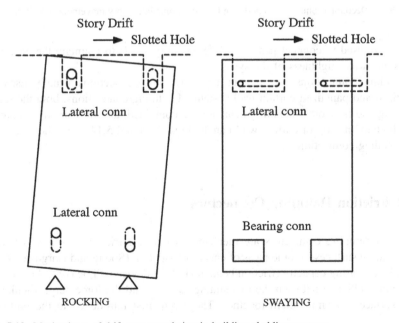

Fig. 5.10 Mechanisms of drift accommodation in building cladding systems

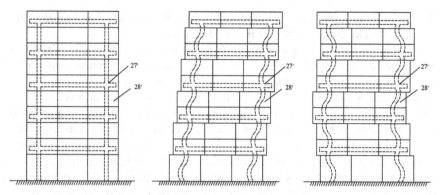

Fig. 5.11 Depiction of earthquake isolated curtain wall system; vertical mullions are attached to building frame at only one story level

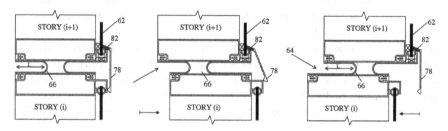

Fig. 5.12 Decoupler joints accommodating building frame inter-story movements (Wulfert 2003)

joints are used to connect panels at different story levels to maintain the air-tightness and water-tightness of the system.

Sliding connections are another type of sway connections usually used in unitized and panelized curtain wall systems. In sliding connections, both the load-bearing and horizontal isolating behavior are combined, and they are used both at the bottom and top of curtain wall panels, Figs. 5.13 and 5.14 show the details of two sliding connections.

5.2 Friction Damping Connectors

Friction damping connectors are mechanisms that use friction—usually between two sliding surfaces—as the basis of energy dissipation (Soong and Dargush 1997). The load-bearing capacity (friction behavior) of the device will be controlled by the friction coefficient between the two sliding surfaces and the force perpendicular to the surfaces attaching them together. They were first introduced to the building

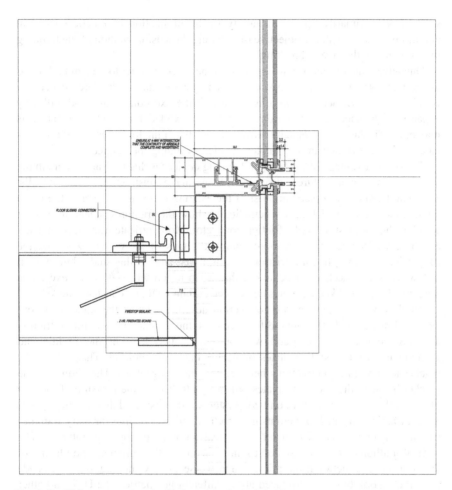

Fig. 5.13 Sliding connection details for unitized curtain wall systems

Fig. 5.14 Horizontal clearance at unitized curtain wall connections

sector based primarily upon an analogy to the dissipation of kinetic energy in automotive brakes with the objective of reducing the seismic motion of the building by "braking" rather than "breaking".

One advantage of this system is that it can be tuned in order to transmit a limited amount of force between the two connected systems, and if the force exceeds a certain limit the device will no longer transmit the exceeding force and will only experience displacement in the direction of the exerted force until it reaches its maximum displacement capacity. How to find the limit force for a certain glass panel is described later in the chapter devoted to tuning the connection.

The other advantage of the friction damping devices is that they are rather simple devices, easy to manufacture and very easy to use in the construction of curtain walls. They can be applied in place of most of the existing brackets with very little or no modification. Figures 5.15a, b schematically show the friction connection bracket which is the simplest form of a friction connector. More intricate friction connectors can be used for different and more complex details of the curtain wall and the supporting structure, like the friction rod discussed later in this study, but they all follow the same mechanical behavior, that is, to let go when the force exceeds a predefined limit and keep the pressure on the curtain wall at a constant rate.

As shown in the figures, the connection device is almost the same as a connection bracket with the only difference that the bracket in use for a friction damping connector is provided with a long slotted hole in the middle and accompanied by two sliding plates on each wing of the bracket. These allow the bracket to slide in two directions between the enclosing plates. The sliding of the bracket between the enclosing plates will be controlled by the pressure of the two plates on the bracket wings. In unitized systems it is considered that the glass panes have a framing attached to them using structural silicon bonding (primary framing) which is again connected to a set of vertical mullions (secondary framing). The vertical mullions are fully fastened to the structure of the building. The placing of the connection is between the primary and the secondary framing systems. Aside from the friction bracket introduced above different geometries like U, Z and other shape brackets can be used as friction connecting devices to connect different framing systems and accommodating isolation in desired directions.

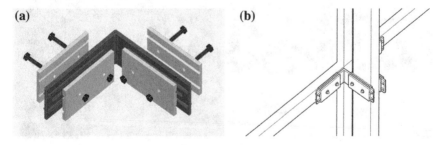

Fig. 5.15 Friction connector bracket. **a** Connection components, **b** Attachment to substructure

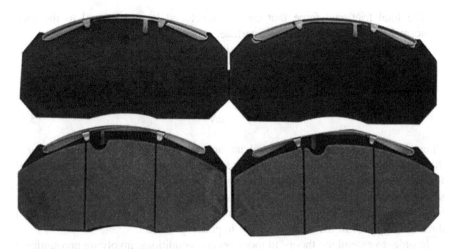

Fig. 5.16 Brake lining pads

It is essential for the materials used in friction connectors to present the two characteristics below:

To have similar friction coefficients in both cases of static and dynamic frictions between the sliding surfaces.

To have considerable resistance against environmental attacks, since corrosion and other changes which may happen to the surface of the plates can dramatically change the behavior of the connection device.

Incorporating brake lining pads that are used in automotive brakes as the enclosing brakes over sand blasted steel plates will result in a consistent force-displacement response and an almost perfect square hysteresis loop while satisfying the conditions above for friction connectors (Fig. 5.16).

The friction coefficient between the brake lining pads and blasted steel disks used in automobile brake systems in most cases is somewhere in-between 0.3 and 0.4.

5.3 Basis and Behavior

The scientific study of solid friction has a long history dating back to da Vinci, Amontons and Coulomb. The basic theory is funded upon the following hypotheses, which were initially inferred from physical experiments with planar sliding of rectangular blocks:

1. The total friction force that can be developed is independent of the apparent surface area of contact.

2. The total frictional force that can be developed is proportional to the total normal force acting across the interface.
3. For the case of sliding with low relative velocities, the total frictional force is independent of that velocity.

As a result of these assumptions we have:

$$F = \mu N \tag{5.1}$$

where F and N represent the frictional and normal forces and, respectively, and μ is the coefficient of friction. Since it is frequently observed that the coefficient of friction is somewhat higher when the slippage is pending than it is during sliding, separate coefficients for static (μ_s) and kinetic (μ_k) frictions are introduced. But in both cases the friction force acts tangential to the interfacial planes and with a direction opposite to the motion, or the impending motion.

In order to extend the theory to more general conditions, involving non-uniform distributions or non-planar surfaces, these basic assumptions are abstracted to infinitesimal limit. Thus total forces are replaced by surface tractions, the generalization of Eq. (5.1) becomes:

$$\tau_t = \mu \tau_n \tag{5.2}$$

in terms of normal traction τ_n and tangential traction τ_t. This form is useful for determining the normal contact stresses that are often required for proper design. The above generalization will be later used for determining the overall characteristic of the friction rod, but for planar surfaces Eq. (5.2) is used as basis of the frictional behavior.

The concept of Coulomb friction, as described above, provides the theoretical basis for most of the work that has been done concerning friction damping. However it should be noted that the Coulomb model is an approximate modeling of the friction and a very good one for simple mechanical problems but does not consider all of the influencing parameters, for example the friction coefficient that is assumed to be constant for a pair of sliding materials may also depend on the surface treatments between the two, and the environmental parameters like temperature. Anyway for the purpose of this research there is no need for the identification of modern theories of friction and the microscopic mechanisms involved with the interfacial bonding and the simple Coulomb theory for infinitesimal surfaces is satisfactory.

References

Bai, B. (2009). Connecting device for curtain wall units, no. 20090249736.

Brueggeman, J. L., Behr, R. A., Wulfert, H., Memari, A. M., & Kremer, P. A. (2000). Dynamic racking performance of an earthquake-isolated curtain wall system. *Earthquake Spectra, 16*(4), 735–756.

Chopra, A. K. (2001). *Dynamics of structures: Theory and applications to earthquake engineering* (2nd ed.). Upper Saddle River: Prentice Hall, Prentice-Hall International.

De Gobbi, A. (2010). *Curtain wall anchor system*, US7681366 ed, US.

FitzGerald, T. F., Anagnos, T., Goodson, M., & Zsutty, T. (1989). Slotted bolted connections in aseismic design for concentrically braced connections. *Earthquake Spectra, 5*(2), 383–391.

Goodno, B. J., Craig, J. I., Dogan, T., & Towashiraporn, P. (1998). *Ductile cladding connection systems for seismic design*. U.S. Department of Commerce, Technology Administration, National Institute of Standards and Technology.

Goodno, B. J., & Zeevaert Wolff, A. (1989). *Working group conclusions on cladding and nonstructural components*.

Goodno, B., Zeevaert-Wolff, A., & Craig, J. I. (1989a). Behavior of heavy cladding components. *Earthquake Spectra, 5*(1), 195–222.

Goodno, B. J., Craig, J. I., & Zeevaert Wolff, A. (1989b). *Behavior of architectural nonstructural components in the Mexico earthquake. Final progress report*.

Memari, A. M., Behr, R. A., & Kremer, P. A. (2003). Seismic behavior of curtain walls containing insulating glass units. *Journal of Architectural Engineering, 9*(2), 70–85.

Pall, A. S., & Marsh, C. (1982). Response of friction damped braced frames. *ASCE Journal of Structuring Division, 108*(ST6), 1313–1323.

Pall, A. S., Marsh, C., & Fazio, P. (1980). Friction joints for seismic control of large panel structures. *Journal Prestressed Concrete Institute, 25*(6), 38–61.

Pantelides, C. P., & Behr, R. A. (1994). Dynamic in-plane racking tests of curtain wall glass elements. *Earthquake Engineering and Structural Dynamics, 23*(2), 211–228.

Pantelides, C., Deschenes, J., & Behr, R. (1993). Dynamic in-plane racking tests of curtain wall glass components. In *Structural Engineering in Natural Hazards Mitigation*, p. 664.

Pantelides, C. P., Truman, K. Z., Behr, R. A., & Belarbi, A. (1996). Development of a loading history for seismic testing of architectural glass in a shop-front wall system. *Engineering Structures, 18*(12), 917–935.

Pinelli, J. P., Craig, J. I., Goodno, B. J. & Cheng-Chieh, H. (1993). Passive control of building response using energy dissipating cladding connections. *Earthquake Spectra, 9*(3), 529–546.

Pinelli, J. P., Craig, J. I., & Goodno, B. J. (1995a). Energy-based seismic design of ductile cladding systems, *Journal of Structural Engineering—ASCE, 121*(3), 567–578.

Pinelli, J. P., Craig, J. I., & Goodno, B. J. (1995b). Energy-based seismic design of ductile cladding systems. *Journal of Structural Engineering, 121*(3), 567–578.

Pinelli, J. P., Moor, C., Craig, J. I., & Goodno, B. J. (1996). Testing of energy dissipating cladding connections. *Earthquake Engineering and Structural Dynamics, 25*(2), 129–147.

Soong, T. T. (1990). *Active structural control: Theory and practice, Longman Scientific & Technical*. New York: Wiley.

Soong, T. T., & Constantinou, M. C. (1994). *Passive and active structural vibration control in civil engineering*. New York: Springer.

Soong, T. T., & Dargush, G. F. (1997). *Passive energy dissipation systems in structural engineering*. New York: Wiley.

Wulfert, H. (2003). Earthquake-immune curtain wall system, no. 6598359.

Chapter 6
Rotational Friction Connection

Abstract One of the major contributions of this study is the introduction of a novel connection device that may be used in complex architectural glazing systems that utilization of the previously discussed connection systems is not applicable. In this newly proposed connection device, the friction mechanism is incorporated between spherical and cylindrical surfaces. In this chapter the development of the idea of the connection device is presented as well as the adjustments necessary for adapting it for different architectural glazing systems. Finally based on the Coulomb theory of friction, the governing equations that control the behavior of the connection device are presented.

6.1 Rotational Friction Connector

In typical friction connection devices, the horizontal loads that are applied due to lateral drifts in the structure are directly the acting forces which result in triggering the friction behavior. In this research, however, an innovative friction device is introduced that does not directly use transversal applied forces on the friction surfaces, but instead utilizes the effects that the horizontal loads will have on the connection as means to trigger the friction mechanism. In this connection device the moments that result from the out-of-plain distance (outward length of the connection) between the applied loads and their reaction points, are used for creating friction energy dissipation, Fig. 6.1. For this reason the connection is called a Rotational Friction Connection (RFC) or the Friction Moment Rod (FMR).

Considering the rotational stiffness for the two ends of the connection the relation between the applied forces and the resulting moments becomes:

$$M_1 + M_2 = FL \rightarrow F = \frac{M_1 + M_2}{L} \tag{6.1}$$

© The Author(s) 2015
R. Afghani Khoraskani, *Advanced Connection Systems for Architectural Glazing*,
PoliMI SpringerBriefs, DOI 10.1007/978-3-319-12997-6_6

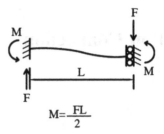

Fig. 6.1 The resulting moments from applied loads at the ends of connection device

where M_1 and M_2 are the moments resulting at the two ends of the connection, one connected to the structure the other to the glazed system. Usually the connection is designed in such a way as to eliminate or minimize the moments applied on the curtain wall system.

Although the energy dissipating ability of the friction connection devices (or any other advanced connection device) may not play a significant role in the dynamic response of the entire building in the case of light-weight cladding systems—as in glass curtain walls—this property can play an essential role in the dynamic behavior of the envelope system itself. The energy dissipation, by minimizing the racking of the envelope system during and after a seismic event, will highly reduce the probability of damage occurring in the envelope system and its attachments.

In this research the friction connection rod is introduced as an innovative connecting element that can almost be incorporated in all of the glazed envelope systems without disturbing their visual characteristics and at the same time protecting them against undesired forces caused by deflections within the actual structure. Two types of rotational friction devices are designed for unitized and suspended glazing systems. The basis of the behavior for both systems is mainly the same but due to geometrical, shape and attachments of these systems adjustments are made for these connections so that they can be easily adopted in the two different systems without much change and modification in the existing elements and attachments of these systems.

6.2 Development of the Idea

The idea of a rotation friction damping device came up from an attempt to satisfy the two main objectives below:

> To introduce a connection device than can handle relatively large displacements in as many degrees of freedom as possible in rotation and translation.
> To maintain the geometrical integrity of the envelope system and manage the connection device to behave in a controllable manner.

Typically to satisfy the first objective, employment of elastic or visco-elastic material seems to be the first candidate for the isolation mechanism, the same as that typically used in the bolting elements of spider glazing systems to eliminate local stresses and reduce local moments imposed on glass cornets, Fig. 6.2. But these mechanisms have the disadvantage of undergoing some undesired and unaccounted for displacements. Such behavior would conflict with the second objective when dealing with relatively large displacements, especially in cases where a moderate level of geometrical complexity is present or when the geometrical assembly tolerances are very low.

On the other hand among different mechanisms that could be adopted for satisfying the second objective mentioned above, the friction mechanism was found to be the most appropriate. This is because basically friction connectors behave the same as rigid connections prior to their slippage and unless the forces and displacement applied on the envelope exceed a damaging limit they act as rigid

Fig. 6.2 Bolted attachment devices using elastic dampers

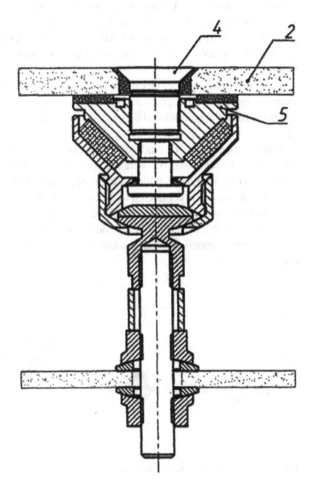

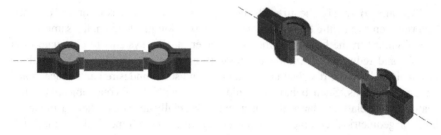

Fig. 6.3 Cylindrical rotational friction connector

connections, hence maximizing the mechanical integrity of the system. But the problem with general types of friction connectors is that they usually can provide only one degree of freedom in a transversal direction, also integrating them in geometrically complex systems is somehow challenging.

So it was decided to collect the advantages of these two isolating mechanisms in a single connection device. Figure 6.3 demonstrates the first draft of the rotation friction device where instead of flat surfaces the friction occurs between two cylindrical surfaces and providing degrees of freedom in both translation and rotation.

Later the cylindrical components of the device were replaced by spherical elements to increase the degrees of freedom in both horizontal and vertical directions and increasing the geometrical adaptability of the device for more complex systems, Fig. 6.4.

Finally, changes were made to the relative dimensions of the two ends of the connection device based on their anchor points. Due to relative delicacy and fragility of envelope systems with respect to their supporting structure, the elements on the side that was to be connected to the structure were considered to be larger in size than the ones connected to the envelope system, Fig. 6.5. Aside from directing the forces towards the larger and stiffer side this made the visual characteristics of the device more compelling and increased the aesthetical values of the device in exposed systems.

Fig. 6.4 Spherical rotational friction connector

Fig. 6.5 The friction moment rod (FMR)

6.3 For Point-Fixed Systems (Friction Moment Rod)

Ever since the introduction of friction damping connections to the building industry, scholars and designers have introduced many devices to connect different structural systems or elements within one structural system. Some were installed between structure trusses, some to connect floor panels and some even to connect heavy concrete claddings to the main structure of the building, and they all have been used with the concept of maximizing energy dissipation within the structure. The forces acting upon them were on the level of structural forces and they were almost always hidden within the structure of the building and not a part of its outlook. Utilizing the friction connectors in a delicate system such as a curtain wall calls for great attention to many other aspects not usually considered during the design of these types of devices. Among these aspects are: aesthetical features of the device, the delicacy of the device and its applicability for different curtain wall manufacturing and construction techniques.

The friction brackets, replacing the common connecting elements for a typical curtain wall system, was already discussed, in general, in this research as one of the most simple friction connecting devices that can be cheaply and easily manufactured and used as curtain wall connecting devices. But it can only be used in a very limited group of glazed envelopes and special conditions for framing geometry. The possibility of benefiting from more geometrically and technically complex envelope systems with the ever improving technical advances within this field, along with the great tendency among architects and designers for adapting them in their buildings, calls for connecting devices compatible with these envelope systems.

One group of envelope systems that has gathered the attention of architects, especially in the trend of modern architecture, is suspended glazing systems. But due to high exposure of these systems and high aesthetical demand over their composing elements the use of common dissipating or isolating connectors is impossible within them, and there exists a lack of proper connection devices that can provide isolation, if required, in these systems. And considering the facts that every individual glass pane in these systems is fully fastened to their supporting structure and that there is no clearance between the composing panels, these systems have became highly susceptible to damage in the event of an earthquake.

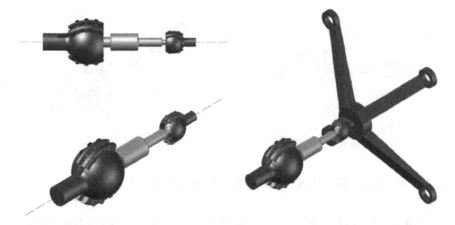

Fig. 6.6 The friction moment rod (FMR)

The friction moment rod is a connection device especially designed for these envelope systems in order to compensate for the lack of proper connection devices for these relatively expensive envelope systems (Fig. 6.6).

The friction rod is consisted of a tubular steel rod, accompanied by two steel spherical balls at its two ends and a pair of supporting emptied spheres, over the steel balls attached to the rod, as shown in Fig. 6.7. The connection of the spherical supports to the glazed envelope or to the structure of the building can vary depending on the requirements and details of the two systems. They can either be welded or screwed to the two systems or they may already be attached to the elements used in the two systems during manufacturing, Fig. 6.7b. The spherical shapes of the two ends of the rod will give us the advantage of providing controllable isolation in four degrees of freedom, which are horizontal and vertical translations in the plane perpendicular to the axis of the rod and rotation over these two directions. This also makes the adjustments of the two systems quite easy during the construction phase.

(a)   **(b)**

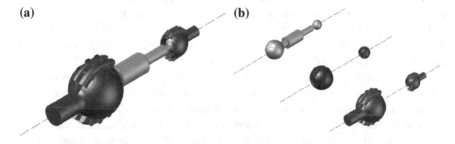

Fig. 6.7 The friction moment rod (FMR) components. **a** Rotational Friction Connector, **b** Its components

As shown in Fig. 6.7 the supporting spheres have an opening gap that divides them in two parts, and a set of bolts is used over the opening gap to control the normal traction applied over the inner sphere that causes the tangential friction traction over the surface of the inner sphere. In order to guarantee a consistent and predictable frictional behavior the inner surface of the supporting spheres are covered with 2 or 3 mm of brake lining material. The forces applied perpendicular to the direction of the middle rod will result in a torque at the two ends of the connecting device which triggers the frictional rotation of the inner sphere inside. These relations are described more in detail below.

6.4   For Unitized and Panelized Systems

Based on the same concept that led to the design of Friction Moment Rod for suspended glazing systems, another rotational friction connection device is also presented in this research which is specially designed for unitized, panelized and stick curtain wall systems. Instead of having sphere shaped balls at the two ends of the FMR, this device has two steel cylindrical attachments at the ends of the connection body which limit the rotation of the device only over a vertical axis, and therefore it can only accommodate mechanical isolation in the horizontal direction. Other than the connection body there are two other composing elements in this connection; one which is anchored to the structure and the other to the curtain wall. The connection of this device to the structure is placed on the concrete floor deck with a steel plate that is embodied in the floor concrete and anchored to it. Above the plate a triangular section attached to semi-cylindrical brackets is situated and the brackets will cover the cylindrical end of the connection body. Between the cylindrical surfaces again a brake lining pad material is used for consistent frictional behavior. A pair of pressure bolts control the pressure applied over the cylindrical tube and control the friction behavior of the connection device. At the other end, the connection body is pinned to a U bracket attached to either vertical or horizontal mullions of the curtain wall. At this point no frictional behavior is expected and the bracket can freely rotate over the vertical axis around the pin. Schematic shape and details of this connection device is presented in Fig. 6.8 and the details of the connection to the structure and the unitized curtain wall is demonstrated in Fig. 6.9.

Unlike the suspended glazing systems, there are many solutions for providing mechanical isolation in unitized glass systems other than the rotational friction connection, mostly discussed in with respect to advanced connection devices in Chap. 5 of this research. But one advantage of the rotational friction connection is that it is highly controllable and unless subjected to damaging forces—damaging to the envelope system—it acts exactly the same as a rigid connection, this will eliminate unaccounted for displacements in the curtain wall system and increase its overall reliability.

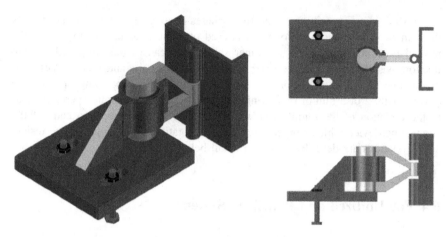

Fig. 6.8 Rotational friction connection for unitized and stick curtain wall systems

6.5 The Behavior of the Rotational Friction Connector

Figure 6.10 demonstrates the schematic behavior of the rotational friction connector for both unitized and suspended glazing systems.

Before slippage occurs between spherical surfaces, the structural model of the connection device is a clamped beam with displacement degree of freedom perpendicular to the axis of the beam as in Fig. 6.11. And the resulting moment over the steel ball equals:

$$M_1 + M_2 = FL \rightarrow F = \frac{M_1 + M_2}{L} \tag{6.2}$$

where F is the force cause by the lateral displacements of connection ends and L is the distance between the connection supports. Even after the interacting surfaces of the spheres begin to slide over each other, this relation will stay valid if the characteristics of the two ends are considered to be the same and the slippage starts to occur in both of the two ends at the same time.

It is now possible to control the lateral force transferred through the connection body by controlling the resulting moment at its two ends.

6.5.1 Formulations for the Friction Moment Rod (Point-Fixed Systems)

On the other hand it can be shown that the torques at the two ends of the rod are results of the friction tractions that happen between the outer surface of the steel ball and the inner surface of the supporting sphere.

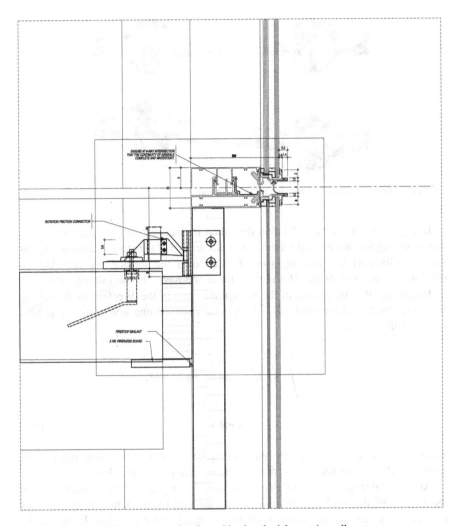

Fig. 6.9 Rotational friction connection for unitized and stick curtain wall systems

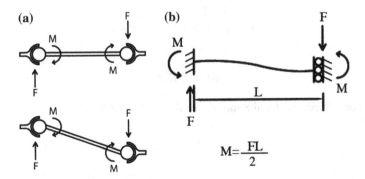

Fig. 6.10 Structural model of the friction moment rod; **a** schematic representation of the rotational friction device, **b** structural model

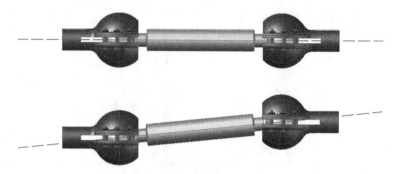

Fig. 6.11 Slipage occurring in the friction moment rod

Using the theory of solid friction described in Sect. 7.2 for non-planar surfaces and assuming that the normal traction over the surface of the steel ball caused by the joining bolts over the support is constant all over the surface, we can find the relation between the torque M and the sum of the forces in the joining bolts.

Integrating the component of the normal tractions in the direction of the joining bolt over one half of the steel ball, which corresponds to one side of the supporting sphere, will result in:

$$N = \int_0^{2\pi} \int_0^{\pi/2} R^2 \tau_n \sin(\theta) \cos(\theta) d\theta d\varphi \qquad (6.3)$$

$$N = \pi R^2 \tau_n \qquad (6.4)$$

where N is the force applied by the joining bolts through the supporting hemi-spheres and τ_n is the normal traction on the surface of the steel ball.

At the same time, integrating the moment of the tangential friction tractions over the surface of the steel sphere gives us the amount of the acting torque M with respect to the tangential friction tractions. It should be noted that the tangential tractions over the surface of the steel ball oppose the direction of the slippage.

$$M = \int_0^{2\pi} \int_0^{\pi} R^3 \tau_t \sin^2(\theta) d\theta d\varphi \qquad (6.5)$$

$$M = \pi^2 R^3 \tau_t \qquad (6.6)$$

having considered that for the infinitesimal surfaces we have:

$$\tau_t = \mu \tau_n.$$

Determining τ_n from Eqs. 5.2 and (6.4) and putting it into Eq. (6.6) will give us the relation between the force applied by the bolt and the torque that it can handle, for each steel ball based on the pressure of the bolts and the radius of the sphere we have:

$$M = \pi \cdot \mu \cdot R \cdot N. \tag{6.7}$$

Having in mind the Eq. (5.3.2.1) relating the lateral force on the friction rod to its torque, we can find the relation between the lateral forces which can be transferred through a friction rod, with respect to the pressing force applied on its ending steel balls with the joining bolts:

$$F_L = \frac{\pi \cdot \mu}{L} \cdot (R_1 N_1 + R_2 N_2). \tag{6.8}$$

6.5.2   Formulations for the Rotational Friction Connection for Unitized Systems

Again using the theory of solid friction described in Sect. 7.2 for non-planar surfaces, and assuming that the normal traction over the surface of the steel ball is constant all over the surface, we can find the relation between the torque M and the sum of the forces in the joining bolts. The only difference here with the formulation of the FMR is that the integrations are applied on cylindrical surfaces instead of spherical ones.

Integrating the component of the normal tractions in the direction of the joining bolt over one half of the steel cylinder, which corresponds to one side of the supporting brackets, will result in:

$$N = \int_0^h \int_0^\pi R\tau_t \sin(\theta) d\theta dz \tag{6.9}$$

$$N = 2Rh\tau_n \tag{6.10}$$

where N is the force applied by the joining bolts through the supporting hemispheres and τ_n is the normal traction on the surface of the steel cylinder. Here h represents the height of the cylindrical steel that is being covered inside the supporting brackets and is in contact with brake lining pads.

At the same time, integrating the moment of the tangential friction tractions over the surface of the steel cylinder gives us the amount of acting torque M with respect to tangential friction tractions. It should be noted that the tangential tractions over the surface of the steel cylinder oppose the direction of the slippage.

$$M = \int_0^h \int_0^{2\pi} R^2 \tau_t d\theta dz \tag{6.11}$$

$$M = 2\pi R^2 h \tau_t. \tag{6.12}$$

Having considered that for the infinitesimal surfaces we have:

$$\tau_t = \mu \tau_n.$$

Determining τ_n from Eqs. (5.2.2) and (5.3.2.3) and putting it into Eq. (5.3.2.5) will give us the relation between the force applied by the bolt and the torque that it can handle:

$$M = \pi \cdot \mu \cdot R \cdot N. \tag{6.13}$$

Having in mind the Eq. (5.3.2.1) relating the lateral force on the friction rod to its torque, we can find the relation between the lateral forces which can be transferred through a friction rod with respect to the pressing force applied on its ending steel balls with the joining bolts:

$$F_L = \frac{\pi \cdot \mu \cdot R}{L} N. \tag{6.14}$$

Adjusting the internal forces of the pressure bolts is the way to control the limit force of the connection device and, now that we have the relation between pressure bolts and the limit force for both of the rotational friction connectors, it is time to set limit force values for different glazing systems and sizes. This will be done in the next chapter dedicated to tuning friction connection devices.

6.6 Friction Lining Material

Having discussed the basis of the behavior of friction damping connectors it is now required to define materials and mechanisms that behave consistent with the theory in order to avoid unforeseen results. Pall et al. (1980) conducted static and dynamic tests on a variety of simple sliding elements having different surface treatments in order to find a system showing a consistent predictable response. Figure 6.12 shows the result of hysteretic behavior under displacement controlled cyclic loading.

As is apparent from the figure the system containing brake lining pads between steel plates provides a consistent predictable response. It is perhaps not surprising

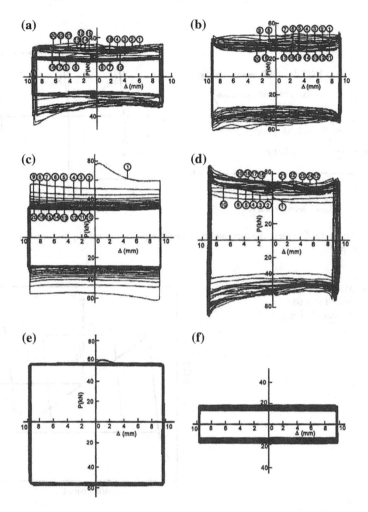

Fig. 6.12 Hysteresis loops of limited slip bolted joints (Pall et al. 1980). **a** Mill scale. **b** Sand blasted. **c** Inorganic zinc-rich paint. **d** Metalized. **e** Brake lining pads. **f** Polyethylene coating

that brake lining materials would perform well in dynamic tests, given that these materials have been specially developed over time in the automotive industry to provide reliable frictional response (Fig. 6.13).

Employment of brake lining material within the friction connector results in the below hysteretic behavior similar to an elastic-perfectly plastic model. Within the geometrical limits of the device the behavior of the friction connector will be demonstrated by Fig. 6.14.

Where F_L is the limit force value of the connection device, Δ_L is the displacement which in each direction the friction connector can move freely and depends on the geometry of the connection and K_0 represents the mechanical stiffness of the connection device in the direction of the applied loads prior to the slippage.

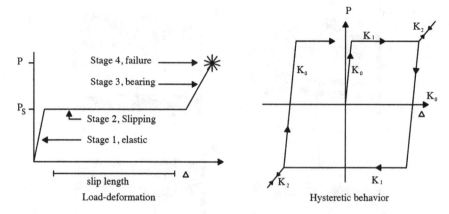

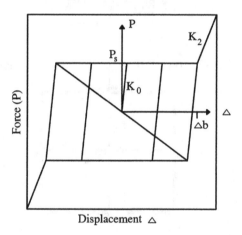

Fig. 6.13 Force-displacement diagram and hysteresis loops for limited slip friction dampers with brake lining pad

Fig. 6.14 Macroscopic model for limited slip friction dampers with brake lining pad

Displacement △

References

Afghani Khoraskani, R. (2012). *Dispositivo Di Collegamento Per Pannelli.* Patent MI2012A000188 ed., Italy.

Ashby, M. F. (2010). *Materials selection in mechanical design* (4th ed.). Oxford: Butterworth-Heinemann.

Bernard, F., Daudeville, L., & Gy, R. (2004). Load bearing capacity of connections in tempered glass structures. *Structural Engineering International: Journal of the International Association for Bridge and Structural Engineering (IABSE), 14*(2), 107–110.

Feng, R., Wu, Y., & Shen, S. (2007). Working mechanism of single-layer cable net supported glass curtain walls. *Advances in Structural Engineering, 10*(2), 183–195.

Feng, R., Yao, B., Wu, Y., & Shen, S. (2010). Dynamic performance of cable net facade with consideration of glass panels under earthquake. *Journal of Harbin Institute of Technology (New Series), 17*(3), 313–317.

Feng, R., Zhang, L., Wu, Y., & Shen, S. (2009). Dynamic performance of cable net facades. *Journal of Constructional Steel Research, 65*(12), 2217–2227.

Pall, A. S., & Marsh, C. (1982). Response of friction damped braced frames. *ASCE Journal of the Structural Division, 108*(ST6), 1313–1323.

Pall, A. S., Marsh, C., & Fazio, P. (1980). Friction joints for seismic control of large panel structures. *Journal—Prestressed Concrete Institute, 25*(6), 38–61.

Sakamoto, I., Itoh, H., & Ohashi, Y. (1984). Proposals for aseismic design method on nonstructural elements. In *Proceedings of 8th World Conference on Earthquake Engineering,* Vol. 5, pp. 1093–1100.

Schlaich, J., Schober, H., & Moschner, T. (2005). Prestressed cable-net facades. *Structural Engineering International: Journal of the International Association for Bridge and Structural Engineering (IABSE), 15*(1), 36–39.

References

Chapter 7
Tuning the Connector

Abstract Providing a desired level of isolation in mechanical terms means to limit the forces and moments acting from one system over the other. These limit forces will be based on the properties of the glazed envelope system and its load-bearing capacity. This evaluation needs to be done in two parts. First within the safety criterion which is to protect the glass panels and its components from failure during or after a severe situation like an earthquake, and second within the serviceability criterion which is, in general, to maintain the desired behavior of the envelope system after the incident in terms of air-tightness and water-tightness and prevent distortions in the envelope surface. To satisfy the safety conditions of a glazed system it is necessary to tune the friction damping device in a way that the forces transferred through it are kept below the maximum allowable forces, which can be handled by the glass panel based on the positions of the connection devices. It is both necessary to study this effect both over a single glass panel and within a group of connected glass panels. An analytical approach is presented for tuning the friction based on the behavior of one-panel window glasses. Then in order to check the results of the analytical solutions a numerical modeling will be performed. With the help of numerical modeling it is also possible to investigate more complex glass panel shapes, the effect of local stresses near the places of the friction damping devices and finally different placement of the connection device over the boundaries of the glass panel.

7.1 Analytical

Usually devices which use the solid friction phenomenon as their dissipative mechanism are tuned in order to maximize the energy dissipation by maximizing the area confined within their hysteresis loop. But the philosophy behind using the friction damping devices in connections of glazed systems to the main structure is

© The Author(s) 2015   85
R. Afghani Khoraskani, *Advanced Connection Systems for Architectural Glazing*,
PoliMI SpringerBriefs, DOI 10.1007/978-3-319-12997-6_7

rather different. These connection devices can be used in order to provide a desired level of isolation between the two systems. Although still energy dissipation will happen within the connection, considering the lightness of these envelope systems and their slenderness with respect to the elements of the main structure, it cannot have a significant effect on the overall dynamic behavior of the structure even if it is tuned for maximized energy dissipation.

What is here referred to as the analytical approach for tuning the connection device is rather a decision making phase in the preliminary design stages of the connection device, rather than a purely analytical solution for tuning the device. At the beginning of a mechanical design process it is necessary to have an understanding of the nature, order and magnitude of the applied forces, in order to make decisions on the dimensions and the behavior of the device that is to be designed.

The results provided in Chap. 4 of this research, on the behavior of glass panels in different glazed envelope systems, are considered as the basis for determining the required information for the analytical study of the forces to design the connection device. Since it was already concluded that the buckling of glass pane is not the main reason of failure in unitized systems, the results for buckling of glass panes in suspended glazing systems will be investigated in this section for FMR connection, and the analysis of the design forces required for friction connections in unitized systems will be left over to the results of experimental tests.

Based on the discussions of Chap. 4 of this research about the behavior of window glass panels during earthquakes, and assuming that the friction connection devices are situated at the four corners of the glass panel, the friction damping devices will be tuned based on the properties of the glass panel. By tuning the friction connector in this section we simply refer to deriving the limit forces to be transferred through the connection device.

Since the lateral displacements in a constant elevation are all of the same magnitude and direction, the forces applied by the friction connectors will be also of the same magnitude (that is, of course, if the connectors are identical). Using the data of critical buckling stresses for glass panels of different sizes and thicknesses it is possible to calculate an estimate for the limit forces of the friction connections. Having known that the horizontal displacements caused by the lateral drifts is the main source of damage to the envelope, only the horizontal component of the applied forces is assumed to be applied by the connection device. Hence, the sum of the limit forces of the connection device on one edge of the panel is assumed to be equal to the sum of horizontal components of the critical buckling forces. The results of the limit forces for two cases of buckling earlier discussed in Chap. 4 are presented in Table 7.1. The values of $F_{L,\tau}$ are the limit force values of the connection based on the distributed shear analysis and the value of $F_{L,P}$ is the limit force value based on diagonal pressure analysis.

These values are purely analytical and have to be evaluated with the help of numerical modeling of the glass panels to provide safety factors for the actual design process.

Table 7.1 Values of tuning forces of friction damping connections for different glass panel sizes and thicknesses

Height (m)	Width (m)	Thickness (cm)	$F_{L,\tau}$ (kg)	$F_{L,P}$ (kg)
1.00	1.00	0.60	128.7492	149.708407
1.00	1.00	1.00	596.0612	693.094476
1.30	0.85	0.60	64.75553	68.9768507
1.30	0.85	1.00	299.7941	319.337272
1.50	1.20	0.60	68.66626	77.8970572
1.50	1.20	0.80	162.7645	184.644876
1.50	1.20	1.2	549.33	623.176458
2.00	1.00	0.8	76.29584	70.9728744
2.00	1.00	1.6	610.3667	567.782995
2.60	1.70	0.8	76.74729	81.7503415
2.60	1.70	1	149.8971	159.668636
2.60	1.70	1.6	613.9784	654.002732

7.2 Numerical

Determining the limit force of the friction connectors with the analytical process earlier described (based on the mechanical strength of the glass panels) is made possible with a group of simplifying assumptions for modeling the behavior of glass panels. The most important of these simplifying assumptions are: having shear buckling as the main cause of failure, uniform distribution of shear force along the edges of the glass panel and considering the glass pane to be part of an imaginary plate and performing the analysis on the imaginary plate. Although these assumptions make analytical modeling possible, we need to make sure they do not put the calculations distant from what happens in reality. One way to achieve more realistic results is to numerically model the behavior of the glazed system under applied loads and with the friction connections. Even though numerical modeling does not have the generality of the closed-form equations of analytical models, they can be used as a tool to evaluate the accuracy and authenticity of them. On the other hand with numerical modeling it is possible to investigate the effect of other expected aspects in the systems that are not represented in the analytical model. Among these aspects are: dynamic nature of the applied forces, nonlinear behavior of the connection device, laminated glass properties, local stresses in the glass panel near the connections and the interaction between adjacent glass panels and its cumulative behavior.

If the results of the numerical and analytical models show considerable differences in value or the behavior of the system under the same loading conditions, it can be concluded that the analytical model does not realistically represent the system in question and we need to generate more realistic models. This is true also in the case that the properties not included in the analytical modeling, like the dynamic nature of the applied forces, show significant effects during the numerical investigation of the system.

The numerical model is divided in two parts, where first the behavior of one panel with the connection devices is investigated and then the behavior of a group of connected panels.

First software, which is used for numerically investigating the system, is SAP2000. In the software the Thin-Shell elements are used for the glass panels and Link elements with an elastic-perfectly plastic behavior in the isolating directions are used for friction connection devices in nonlinear analysis. The modeling of the supporting structure is according to Eurocode 8 (British Standards Institution 1996) when needed. Along with SAP2000 which is a general mechanical simulation software in structural engineering the SJ-MEPLA software is also used which is a commercial structural analysis developed specifically for glass structures and glass components of structural systems.

Considering the fact that the newly introduced connection device named FMR (Friction Moment Rod) is highly suitable for exposed bolted glass systems and in order to reduce the effect of simplifications added to the model a curtain wall system of suspended glazing with bolted assembly (spider glazing) has been considered for the numerical simulations and experimental tests. This decision has been made under the influence of the fact that due to the somehow uniform texture of these systems the results of the numerical simulations will more likely be in close proximity to the real-time behavior of the glazing system. Also a great lack of study in the behavior of the spider glazing systems during and after earthquakes has provided sufficient motive to focus the attention on these glass envelope systems.

7.2.1 Behavior Over One Panel

The first step towards numerically analyzing the earthquake effect on a glass panel is to neglect the effect of other adjacent panels and concentrate on one individual glass pane subjected to forces that are caused by building drifts and applied through connection devices. Although to represent the true nature of the problem a dynamic simulation is required but first we will evaluate the analytical procedures introduced above as the basis for tuning the friction connection devices. In order to evaluate the analytical model, in this part at first a single glass pane is subjected to a set of static forces, representing the limit forces of the friction connectors, in a linear static model.

Later a single panel with a set of connection devices connected to an arbitrary structure is modeled and analyzed using Ritz-vector nonlinear time-history analysis. The arbitrary structure is modeled in a way to show the allowable relative displacements similar to what occurs along a building story during an earthquake.

7.2.1.1 Static Numerical Simulation with SAP2000

At this stage, as described before, a single glass pane is subjected to static loads which are applied through the connections of the glass panel to the structure of the envelope system. Figure 7.1 is a schematic structural demonstration of the simulation.

As shown above all the four resting points of the glass pane are provided with simply supported supports, resisting displacements in the normal direction to the surface of the glass pane, but allowing rotation over the supporting nodes for glass elements in all three dimensions, this is to avoid unnecessary stresses on the glass and mitigate the real-time behavior of the system. In addition, the two bottom supports resist vertical and transversal movements of the glass ensuring structural stability of the numerical model, which is of course a necessity for the simulation. The effects of the transmitted lateral movements of structure are expressed with a set of two acting forces in the transversal direction, over the upper resting points of the glass pane. Again, in order to avoid singularity of forces and unwanted local stress fields in the glass, both the forces and the supports are distributed over the element nodes that correspond to the edges of the bolt holes in the glass pane.

In order to produce an accurate simulation of the stress fields and being able to generate a smooth deformation field in the glass pane, 4 node *thin shell elements* with a maximum dimension of 2.5 cm in every direction are selected for the finite element analysis of the panel.

As for the mechanical characteristics of the glass, due to the common practice tempered glass and laminated glass as the components of the suspended glazing systems—due to post-failure safety reasons—the properties of these glass materials have been considered for the simulations, but as it turned out the elastic and pre-failure properties of these two types of safety glass materials are somehow the same as the monolithic glass and their major difference is related to the post-failure features and their shattering modes. Table 7.2 shows the mechanical properties typically considered for glass materials which are used in the simulations. These data have been extracted from material library of the CES database (Ashby 2010).

Finally a buckling analysis is performed on the glass panes subjected to static forces and the results of the critical buckling forces derived from the SAP2000 software are presented in Table 7.3. The near buckling Von Mises stress state within the glass pane surface and the buckling deformation diagram is presented in

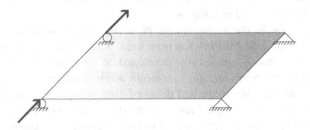

Fig. 7.1 Schematic demonstration of the numerical modeling

Table 7.2 Glass table of properties

Mechanical property	Symbol	Unit	Value
Mass per unit volume	M	kg/m^3	2,400
Modulus of elasticity (young's modulus)	E	Pa	7E+10
Poisson's ratio	υ	–	0.2
Shear modulus	G	Pa	2.9E+10

Table 7.3 Critical buckling horizontal loads for glass panes (SAP2000)

Height (m)	Width (m)	Thickness (cm)	F$_h$ (kg)
1.00	1.00	0.60	270
1.00	1.00	1.00	1,200
1.30	0.85	0.60	115
1.30	0.85	1.00	515
1.50	1.20	0.60	120
1.50	1.20	0.80	270
1.50	1.20	1.2	910
2.00	1.00	0.8	110
2.00	1.00	1.6	880
2.60	1.70	0.8	110
2.60	1.70	1	210
2.60	1.70	1.6	860

Figs. 7.2, 7.3, 7.4, 7.5, 7.6, 7.7, 7.8, 7.9, 7.10, 7.11, 7.12 and 7.13, in order to make comparisons with the results of the SJ-MEPLA software.

7.2.1.2 Static Numerical Simulation with SJ-MEPLA

Although being a very powerful mechanical analysis software for nonlinear static and dynamic simulations, the SAP2000 software does not have the ability to model sandwich shell or plate elements, and for including the characteristics of laminated glass a bending stiffness reduction equal to 0.3 was inserted in the mechanical properties of the glass shell elements in SAP2000. This reduction on the stiffness of the shell elements is based on the corrections in the thickness of laminated glass elements, earlier discussed in Chap. 4.

In order to double check the results of SAP, another numerical simulating software by the name SJ-MEPLA has been used. SJ-MEPLA is commercial finite element software that is specifically developed to perform static and dynamic analysis for glass structures and glass elements within a structure. With MEPLA it is possible to simulate any shape and detail of glass components including; laminated glass, point glass fixings, framing supports around glass panels etc. Every small detail that is within the glass pane or its typical connections can be included in the simulations by MEPLA.

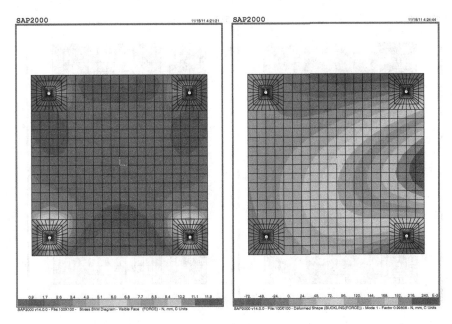

Fig. 7.2 Von Mises stress state and buckling deformation diagram (panel $100 \times 100 \times 0.6$)

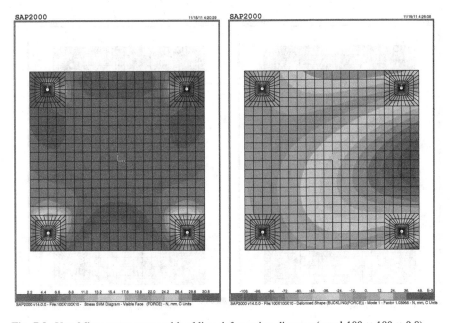

Fig. 7.3 Von Mises stress state and buckling deformation diagram (panel $100 \times 100 \times 0.8$)

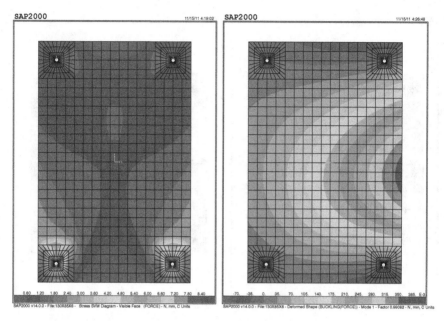

Fig. 7.4 Von Mises stress state and buckling deformation diagram (panel 130 × 85 × 0.6)

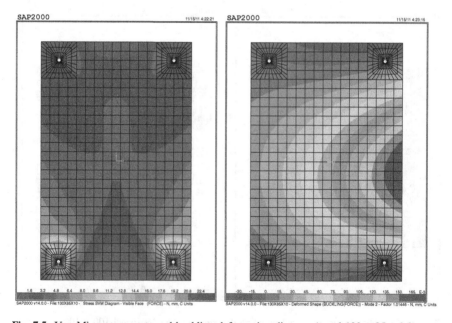

Fig. 7.5 Von Mises stress state and buckling deformation diagram (panel 130 × 85 × 0.8)

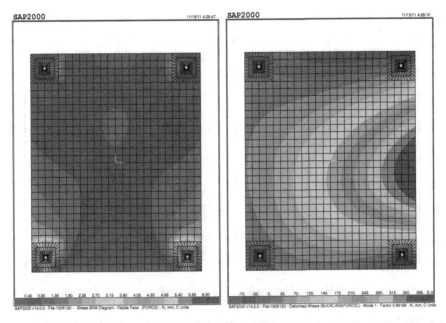

Fig. 7.6 Von Mises stress state and buckling deformation diagram (panel 150 × 120 × 0.6)

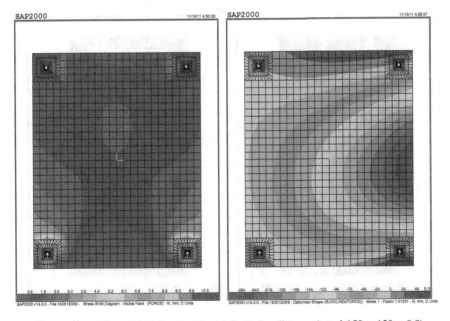

Fig. 7.7 Von Mises stress state and buckling deformation diagram (panel 150 × 120 × 0.8)

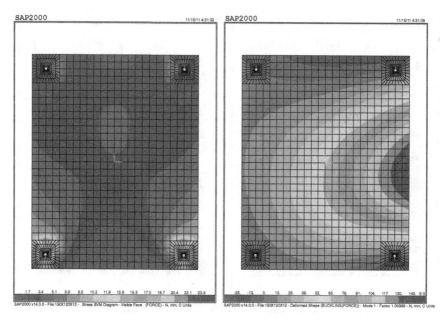

Fig. 7.8 Von Mises stress state and buckling deformation diagram (panel 150 × 120 × 1.2)

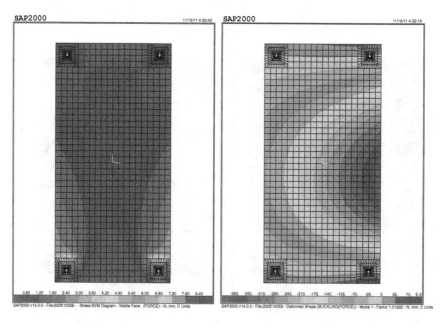

Fig. 7.9 Von Mises stress state and buckling deformation diagram (panel 200 × 100 × 0.8)

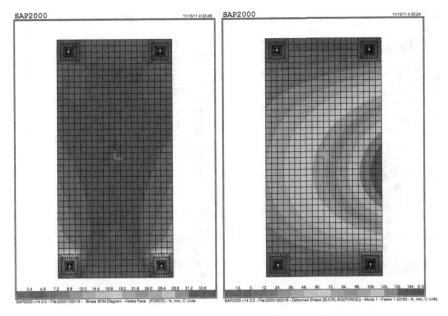

Fig. 7.10 Von Mises stress state and buckling deformation diagram (panel $200 \times 100 \times 1.6$)

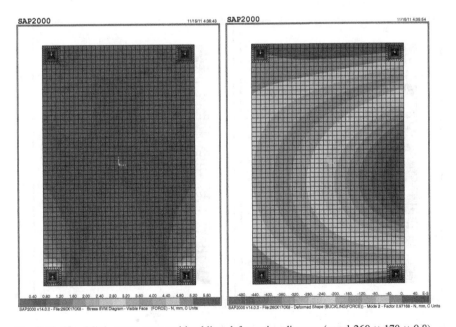

Fig. 7.11 Von Mises stress state and buckling deformation diagram (panel $260 \times 170 \times 0.8$)

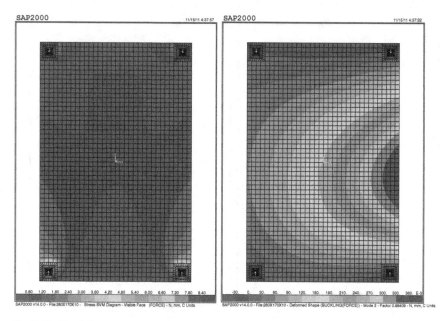

Fig. 7.12 Von Mises stress state and buckling deformation diagram (panel 260 × 170 × 1.0)

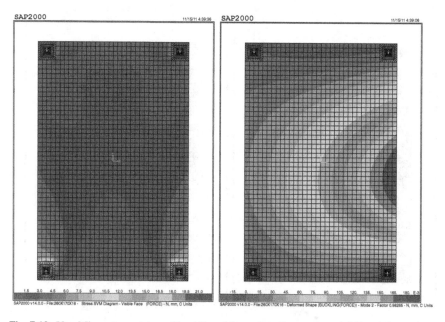

Fig. 7.13 Von Mises stress state and buckling deformation diagram (panel 260 × 170 × 1.6)

As mentioned earlier, SJ-MEPLA is software that performs the calculations based on a finite element model. Contrary to SAP which uses *4 node shell elements*, in MEPLA *9 node shell elements* are used for finite elements, so it is possible to have a mesh composed of bigger elements than the elements of an SAP file but with the same accuracy. The *9 node shell element* can be demonstrated as a rectangle having second-order parabolic curves instead of straight lines between the corner edges, Fig. 7.14.

Using an automatic mesh producing algorithm the MEPLA will divide the glass pane into appropriate shell elements. It is only necessary to indicate the maximum element dimensions. Based on suggestions in the software the maximum shell dimensions vary between 12 and 20 cm. For different glass dimensions, smaller shell dimensions will be used automatically by the software where they are needed, for example near bolted corners of the glass, Fig. 7.15.

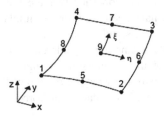

Fig. 7.14 The nine node finite element used in SJ-MEPLA

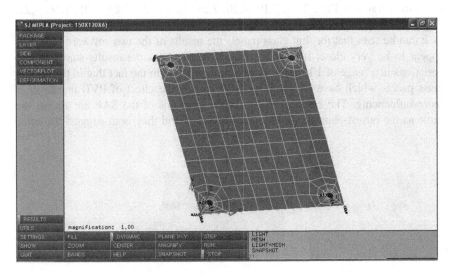

Fig. 7.15 The automatic meshing of the glass panel

Table 7.4 Laminated glass
thicknesses used in SJ-
MEPLA

Glass thickness (mm)	t_{glass} (mm)	t_{PVB} (mm)
6	3	0.38
8	4	0.38
10	5	0.38
12	6	0.76
16	8	0.76

Inputs to MEPLA:

The geometry of the glass panes investigated in MEPLA are the same as the ones earlier modeled in SAP, but the thicknesses of the glass panes are associated in a layered system. For every thickness of the glass pane, Table 7.4 shows the details of the laminated glass used in MEPLA software (Fig. 7.16).

The material properties associated to glass and the PVB interlayer are the same as in SAP simulations. For the properties that are not present in SAP files the suggestions of the software have been used for the simulations. Table 7.5 shows the values of the properties used in simulations.

The bolted fixings of glass pane are considered to be countersunk fixings both they and the dimensions of the countersunk fixings used in the simulations are presented in Fig. 7.17. The two fixings at the bottom of the glass pane are assumed to be fixed to the supports, while the damaging forces will be applied on the top fixings of the glass pane.

The maximum bearing forces on the glass panes which result in the buckling of glass, based on the outputs of MEPLA are presented in Table 7.6, also the stress state and the out-of-plane deformations of the glass panes near the buckling state are demonstrated in Figs. 7.18, 7.19, 7.20, 7.21, 7.22, 7.23, 7.24, 7.25, 7.26, 7.27, 7.28, 7.29 and 7.30.

It can be seen that for thin glass panels the results of the two software packages appear to be very close, but for the thicker glass panes the results start to show variations in a range of 10–20 %. This is probably due to the fact that in the thicker glass panels which have an interlayer of 0.76 mm the effect of PVB interlayer is more influencing. The buckling deformation diagrams of the SAP are almost the same as the out-of-plain deformations in MEPLA and they both suggest the same

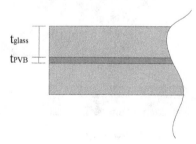

t_{glass}

t_{PVB}

Fig. 7.16 Laminated glass

Table 7.5 Table of properties for glass and PVB (SJ-MEPLA)

Mechanical property		Symbol	Unit	Value
Glass	Mass per unit volume	M_G	kg/m³	2,400
	Modulus of elasticity (young's modulus)	E_G	Pa	7E+10
	Poisson's ratio	υ	–	0.2
PVB	Mass per unit volume	M_{PVB}	kg/m³	1,000
	Modulus of elasticity (young's modulus)	E_{PVB}	Pa	3E+4
	Poisson's ratio	υ	–	0.5

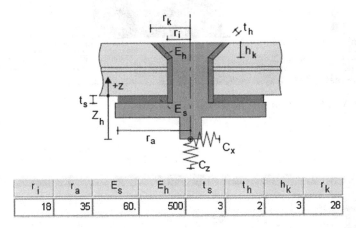

r_i	r_a	E_s	E_h	t_s	t_h	h_k	r_k
18	35	60.	500	3	2	3	28

Fig. 7.17 Details of the countersunk fixing supports used in the simulations

Table 7.6 Critical buckling horizontal loads for glass panes (SJ-MEPLA)

Height (m)	Width (m)	Thickness (cm)	F_h (kg)
1.00	1.00	0.60	270
1.00	1.00	1.00	900
1.30	0.85	0.60	130
1.30	0.85	1.00	450
1.50	1.20	0.60	120
1.50	1.20	0.80	250
1.50	1.20	1.2	700
2.00	1.00	0.8	110
2.00	1.00	1.6	750
2.60	1.70	0.8	110
2.60	1.70	1	200
2.60	1.70	1.6	740

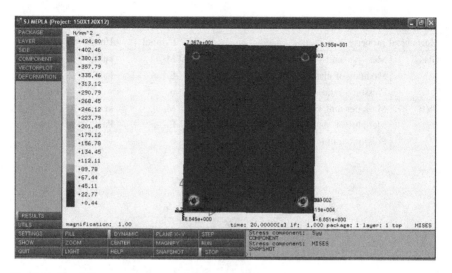

Fig. 7.18 Von Mises stress state (panel 150 × 120 × 0.6)

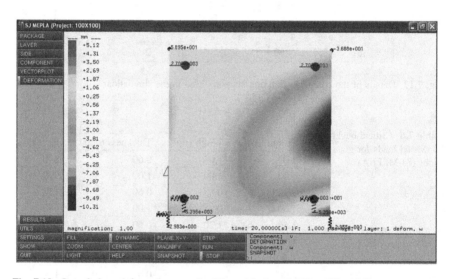

Fig. 7.19 Out of plane deformation near buckling state (panel 100 × 100 × 0.6)

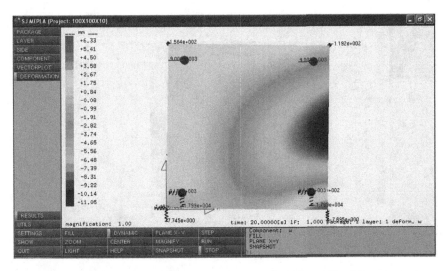

Fig. 7.20 Out of plate deformation near buckling state (panel $100 \times 100 \times 0.8$)

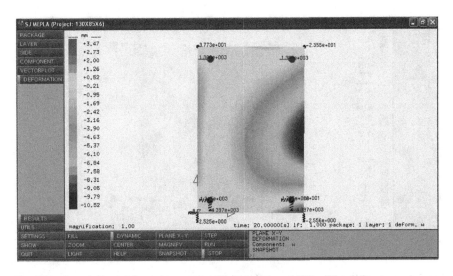

Fig. 7.21 Out of plane deformation near buckling state (panel $130 \times 85 \times 0.6$)

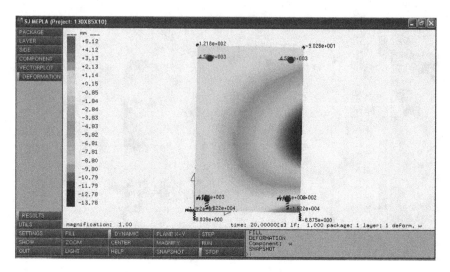

Fig. 7.22 Out of plane deformation near buckling state (panel 130 × 85 × 0.8)

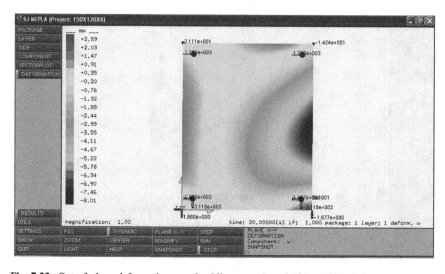

Fig. 7.23 Out of plane deformation near buckling state (panel 120 × 150 × 0.6)

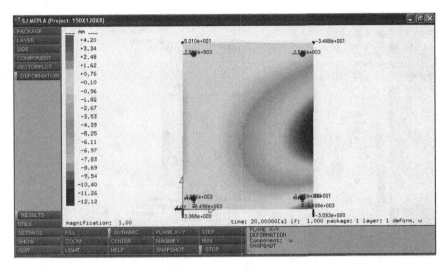

Fig. 7.24 Out of plane deformation near buckling state (panel $120 \times 150 \times 0.8$)

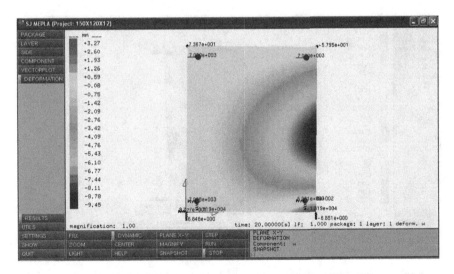

Fig. 7.25 Out of plane deformation near buckling state (panel $120 \times 150 \times 1.2$)

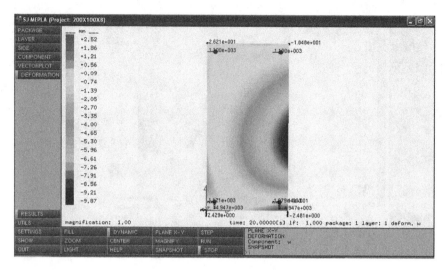

Fig. 7.26 Out of plane deformation near buckling state (panel 200 × 100 × 0.8)

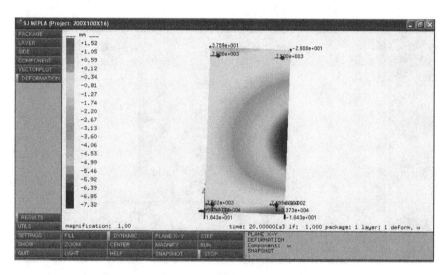

Fig. 7.27 Out of plane deformation near buckling state (panel 200 × 100 × 1.2)

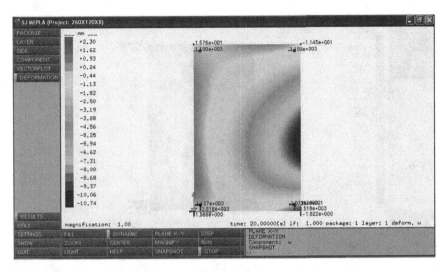

Fig. 7.28 Out of plane deformation near buckling state (panel 260 × 170 × 0.8)

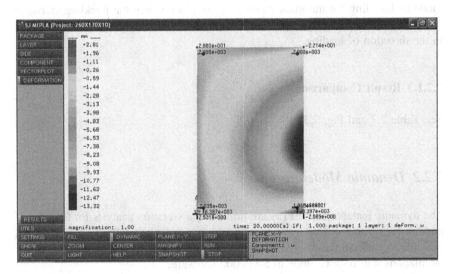

Fig. 7.29 Out of plane deformation near buckling state (panel 260 × 170 × 1.0)

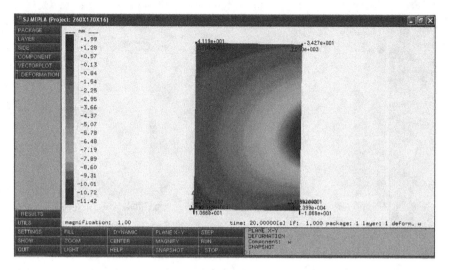

Fig. 7.30 Out of plane deformation near buckling state (panel 260 × 170 × 1.6)

mode of buckling for the glass panes and the place where the buckling starts. In both cases the buckling will start a little below the middle of the vertical edge that is in the direction of loading.

7.2.1.3 Result Comparison

See Table 7.7 and Fig. 7.31.

7.2.2 Dynamic Modeling

The dynamic features of MEPLA are limited to a vibration analysis for the case of pendulum impact—a load case for determining the glass fallout—and it is not possible to perform the desired dynamic analysis with MEPLA, thus the dynamic simulations will only be done in SAP2000 software.

All the previous numerical modelings imposed forces at the connection points on the glass pane, but in reality the forces that are applied on the glass are the product of displacements that occur in the structure during an earthquake. So in the dynamic modeling of the systems, instead of having the glass pane subjected to acting forces, displacements are assigned to the glass supports and a time-history dynamic non-linear analysis is performed. Since the imposed displacements are assumed to be equivalent to the lateral movements in the structure, they are considered to be

Table 7.7 Result of the four approaches considered for analyzing the mechanical buckling of glass panes

Glass pane dimensions	Shear buckling scenario (kg)	Diagonal pressure scenario (kg)	SJ-MEPLA (kg)	SAP2000 (kg)
100 × 100 × 0.6	128.75	149.71	270	270
100 × 100 × 0.8	596.06	693.09	900	1,200
130 × 85 × 0.6	64.76	68.98	130	115
130 × 85 × 0.8	299.79	319.34	450	515
150 × 120 × 0.6	68.67	77.90	120	120
150 × 120 × 0.8	162.76	184.64	250	270
150 × 120 × 1.2	549.33	623.18	700	910
200 × 100 × 0.8	76.30	70.97	110	110
200 × 100 × 1.6	610.37	567.78	750	880
260 × 170 × 0.8	76.75	81.75	110	110
260 × 170 × 1.0	149.90	159.67	200	210
260 × 170 × 1.6	613.98	654.00	740	860

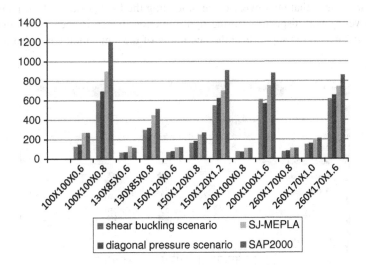

Fig. 7.31 Comparison chart

governed by a sinusoidal function of time which best describes the lateral drifts in the structure during earthquakes. Although the time period of the sine function strongly depends on the building natural frequency and its natural period, a time period equal to 1 s, which is close to natural vibration period of regular midrise buildings, is considered for the sine function. The maximum allowable drift in the structural design of the buildings, in most building codes, is considered to be equal to 0.02, but in this research a drift value of 0.01 has been considered sufficient to observe the dynamic behavior of the structure. So the magnitude of the sine function for loading displacements is considered to be 0.01 multiplied by the height of the glass and three periods of loading is imposed on the glass for every simulation, Fig. 7.32.

At this point two cases of connection to the main structure are considered; first when there are rigid elements connecting the glass to the supports and second when Friction Moment Rod (FMR) is placed between the glass and the supports.

In order to simulate the FMR in the model, a Link element with multi-linear deformation with a perfectly plastic limit equal to the limit force of the FMR based on the buckling analysis is used in place of the connections, Fig. 7.33.

Figure 7.34 shows the stress states at the peak of horizontal displacements of the supports, in the two conditions of using rigid connections and FMR. Animations from the dynamic simulation on glass panels can be found in the CD attached to this document containing numerical simulation files and other outputs that cannot be imprinted in the report.

Comparing the legend of the two figures above it is obvious that glass panel without connection devices has far passed the stresses that can be handled by glass materials. The lateral displacement of the top left corner of the two cases during and after loading is also compared in Figs. 7.35 and 7.36.

It can be seen that after every cycle of loading the final position of the panel will have a very small disposition to the right direction, which is because the imposed displacements on the supports at the beginning of the simulation where from left to

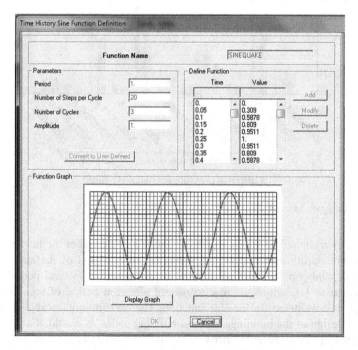

Fig. 7.32 Sinusoidal displacement function representing the lateral drifts of the main structure

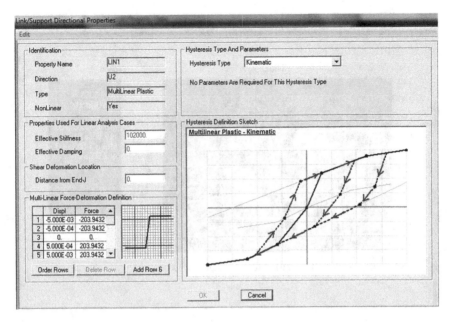

Fig. 7.33 Elastic-perfectly plastic link elements properties for connecting the glass panel to the structure

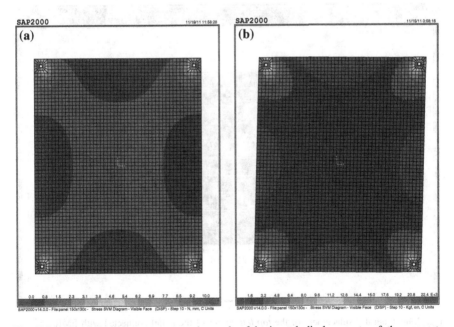

Fig. 7.34 Von Mises stress states at the peak of horizontal displacements of the supports **a** connected with FMR; **b** connected with rigid connections

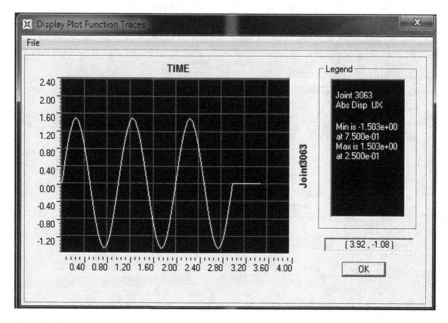

Fig. 7.35 Lateral displacement of the *top left* corner of the panel connected with rigid connections

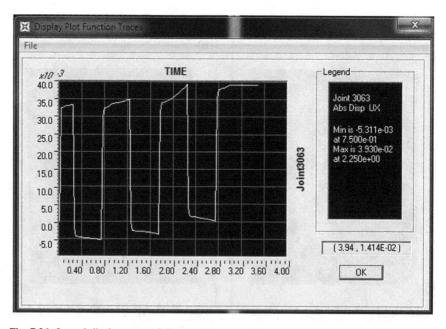

Fig. 7.36 Lateral displacement of the *top left* corner of the panel connected with FMR

right the same effect but in the opposite direction if the initial imposed displacements where from right to left. The reason of this disposition is that after every cycle of loading the friction connector will not return to the zero stress condition and as soon as the applied forces become less than the limit force of the connection, the connection device will become like a rigid connection and at that point the remaining stresses and displacements will be kept in the system; these remaining stresses will also be present in the glass panes after the cyclic loading.

7.2.3 Group of Connected Panels

Having investigated the effects of lateral forces and displacements over a single glass panel, it is time to observe the effects of the lateral drifts over a group of adjacent glass panels. To achieve this, a matrix of glass panels is modeled in SAP, Fig. 7.37.

Neighboring edges of the glass panes is filled with silicon elastonomer material with a width equal to 1 cm and a depth equal to the thickness of the glass panes. To have the maximum possible consistency between the model and a real spider glazing system, the nodes associated to the neighboring holes of the glass fixings, which are supposed to be connected to the same spider bracket, are constrained to undergo movements as a rigid body (meaning no relative displacements will happen between them). This will increase the overall reliability of the system, but decreases the accuracy of the results related to the silicon patches of the corners of the glass and between glass fixings. All the connections used in these simulations are FMR connections modeled Link elements discussed in the former section. Again the dynamic effect of the lateral displacements is imposed with a sinusoidal function with a period of 1 s., the magnitude of the sine function for every support differs based on its elevation and is equal to the vertical distance between the support and the bottom of the model multiplied by 0.01, which is the anticipated drift ratio.

The results of modeling a group of connected panels can be used for two cases below:

In the first case we investigate the cumulative effects that the neighboring panels might have on each other. Aside from the loads applied on the glass pane by the fixings, due to displacements in the structure, the stress state occurring in the glass panel is also related to the forces applied by other panels connected to it. This can be caused by the relative displacements or rotations that may occur between the connected glass panels. Figure 7.38 shows the Von Mises stress state of the glass panel at the peak of the horizontal displacement. Considerable effects of this type were not observed during the simulations of a set of panels and only a small increase to the stress values of the glass panes at the bottom of the set was detected. This effect needs to be further investigated in future research.

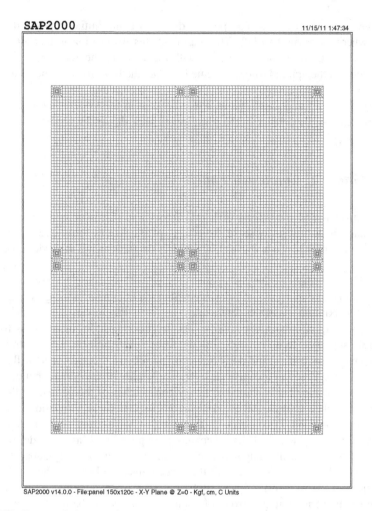

SAP2000 v14.0.0 - File:panel 150x120c - X-Y Plane @ Z=0 - Kgf, cm, C Units

Fig. 7.37 Connected glass panels modeled in SAP

In the second case, we investigate the displacements and stresses in the silicon elastonomer, filling the spaces between the glass panels. With this approach it will be possible to obtain information on whether the silicon elastonomer will maintain its air-tightness and water-tightness properties after a seismic event. There are three points of interest for such observations in the system:

1. In the middle of two vertical edges
2. In the middle of two horizontal edges
3. In between the glass fixings

Figure 7.39 shows the absolute relative displacements between two sides of the silicon patch, each connected to a different glass panel.

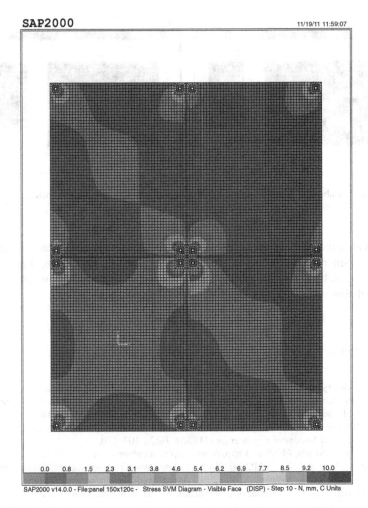

Fig. 7.38 Von Mises stress states at the peak of horizontal displacements of the supports

The safety criterion has been considered in order to derive the limit force values of the friction connector. It is clear that when the safety of the façade system has been assured, the serviceability of the system after earthquake needs to be further investigated. Air tightness and water tightness of the façade are the main features compromised during an earthquake, which might be caused by breaking of the gaskets and failure in the structural silicon. Depending on the details of the façade system, maintaining the functionality of the façade may ask for further reduction in the limit force values of the friction connectors.

It must be noted that a great number of technical details are added to the system when the behavior of a set of connected panels (a mock-up) is under study, and it is not possible to include all these small details within a numerical simulation.

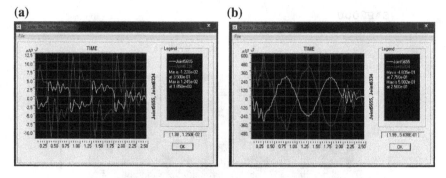

Fig. 7.39 The absolute relative displacements between two sides of the silicon patch **a** connected with FMR; **b** connected with rigid connections

So although the results of simulating a group of connected panels can be used for having a better understanding of the overall behavior of the system, they are not considered to be highly reliable and experimental studies are very much required for validating and evaluating this part of the numerical simulations.

References

American Society of Civil Engineers. (2010). Minimum design loads for buildings and other structures. American Society of Civil Engineers, Structural Engineering Institute, Reston, VA.

Bernard, F., Daudeville, L., & Gy, R. (2004). Load bearing capacity of connections in tempered glass structures. *Structural Engineering International: Journal of the International Association for Bridge and Structural Engineering (IABSE), 14*(2), 107–110.

Bohmann, D. SJ Mepla, (3.5.9 ed.) http://www.mepla.eu/en/home_en.

Computers & Structures Inc. SAP2000, (14th ed.) http://www.csiamerica.com/products/sap2000.

Vyzantiadou, M. A., & Avdelas, A. V. (2004). Point fixed glazing systems: Technological and morphological aspects. *Journal of Constructional Steel Research, 60*(8), 1227–1240.

Chapter 8
Conclusions and Recommendations on Experimental Tests

Abstract Although experimental studies have not been performed within the scope of this research, in this chapter a brief discussion is presented on the experimental tests that are suggested for studying the behavior of the proposed connection devices and their effect on curtain wall systems during seismic events. The aim of these tests will be first to study the behavior of the connection device itself and later to see the effect that it will have on envelope systems subjected to seismic actions. In order to study the mechanical behavior of the connection device a laboratory test apparatus is discussed that is especially conceived for the study of cladding connections. And for the study of the effects of the connection device based on the recommendations of American Architectural Manufacturing Associations (AAMA) (Association of American Architectural Manufacturers 2001), a test facility is proposed to perform two types of static and dynamic experiments on a mockup.

8.1 Behavior of the Connection Device

The theoretical behavior of the rotational friction connectors, presented in this research, was earlier discussed in Sect. 5.3 of this research. In order to make sure the relations between the parameters of the system is realistically achieved, it is necessary to have the connection devices subjected to lateral displacements, such as will happen when connecting the curtain wall to the building structure, and to measure the parameters for limit force values and energy dissipation, in accordance with different values for the applied pressure by the pressure bolts. The laboratory testing machines especially developed for connections in the cladding systems, in earlier works by Pinelli et al. (1996) is described here for running the tests on FMR. The primary objective behind development of the test apparatus was simulation of the behavior of an advanced cladding connector subjected to inter-story drifts, and to that end the machine must have:

© The Author(s) 2015
R. Afghani Khoraskani, *Advanced Connection Systems for Architectural Glazing*,
PoliMI SpringerBriefs, DOI 10.1007/978-3-319-12997-6_8

The ability to isolate and monitor the behavior of the connector elements
The ability to reproduce the actual service loads and deformations to which a connector is subjected during an earthquake
The ability to accommodate different conditions of fixity for the connector ends

In order to achieve these objectives, the machine must be capable of applying a number of specific types of loads to a connector. It is clear that the connector should be arranged in the machine in the same orientation that it would be used to support a vertical curtain wall. In this case the loads and constraints on the connection are:

Horizontal forces or displacements
Moment fixity at both ends
Shear release in vertical direction
Axial release (normal to the vertical façade plane)
Gravity loads if required

The horizontal forces or displacements on the connection device are provided with hydraulic actuators and the gravity load, if needed, is applied by putting weight on the connection. Figure 8.1 provides a schematic figure of the test apparatus and its fixture which is composed of the components below:

A building anchor, which is a thick vertical steel plate attaches to a rigid box, the axial degree of freedom is released with the help of roller bearings in axial direction under the rigid box.
The panel anchor, which also consists of a thick plate connected to a rigid support with long rods in order to prevent axial movements but allow movements in horizontal and vertical directions.
Horizontal load applying system, which is one or two hydraulic actuators attached to the edge of the panel anchor.
Hand operated spring actuators attached to the bottom edge of the panel anchor to apply gravity loads.

More complete details of the test apparatus showing the kinematics of the test fixture are presented in Fig. 8.2.

The testing procedure is that cycles of increasing displacement are applied in small step increments and at each step displacements and forces are recorded and corresponding hysteresis cycles—horizontal forces versus transverse displacements—are plotted. The objectives of the test may include the following:

Evaluate stiffness, ductility and energy dissipation characteristics of the connection.
Evaluate the relations between pressure bolt forces and the behavior of the connection.
Evaluate the consistency of the behavior of the connection after multiple cycles of loading.
Investigate the effect of vertical loading on the connection.

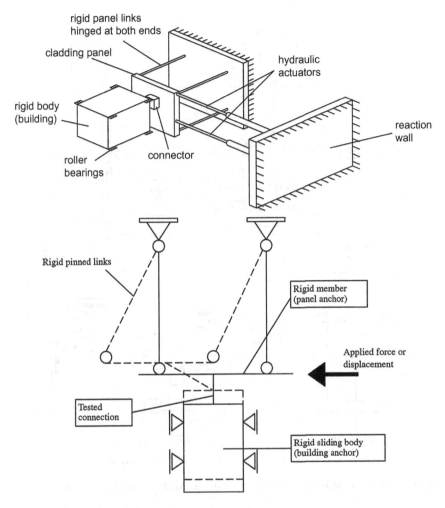

Fig. 8.1 Overall schematic figure of connection test fixture

8.2 Test Mockup

For evaluating the seismic behavior of a spider glazed curtain wall with connection device and subjected to seismic forces, and also to compare the results with the case that there are no friction connectors attached, a test facility is presented in this section which is based on the recommendation of AAMA documents 501.4 and 501.6 for evaluating curtain wall and storefront systems subjected to seismic and wind induced inter-story drifts. Two separate test procedures are considered to be performed on the system: a dynamic racking test for determining the ultimate seismic state of architectural glass and a statically applied load test with primary focus on serviceability of curtain wall system specimens. Same test facility is

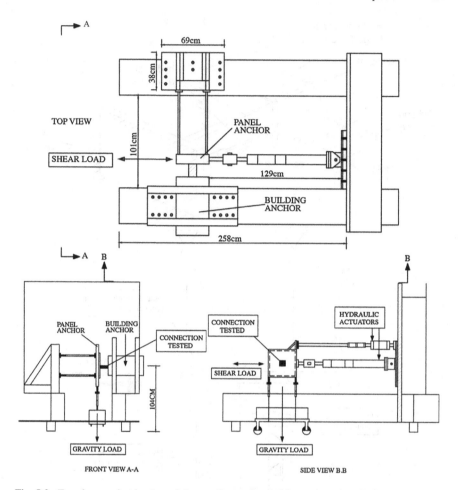

Fig. 8.2 *Top, front* and *side view* of the test fixture for cladding connection devices

proposed for both test procedures which is composed of a spider curtain wall system with full size specimens and components using the same materials, type of glass, details, method of construction and anchorage as those used in actual building, supported by a test chamber structure that simulates the main structural supports of the actual building. However the test chamber support structure may differ from the actual building as it is required to impose the required displacements. The test procedures will differ in static and dynamic cases. These procedures are directly quoted from AAMA 501.4 for static and AAMA 501.6 for dynamic tests (Association of American Architectural Manufacturers 2000, 2001).

The main part of the test chamber structure for simulating the seismic behavior of the actual building is composed of a primary frame with beams and columns hinged for providing required lateral drift. Secondary beams are attached to the main frame of the test apparatus, providing supports for connection elements, and a

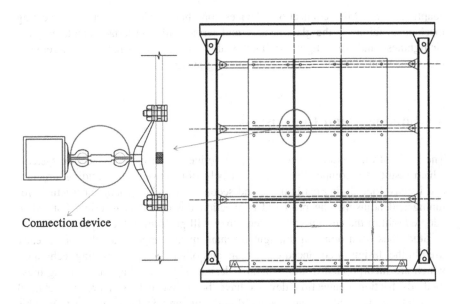

Connection device

Fig. 8.3 Test frame and curtain wall mockup

hydraulic actuator is attached to the top corner of the main frame for controlling the lateral displacements required for the test. A full size spider glazed curtain wall is attached to the secondary beams of the test chamber. The test will be repeated in both cases of adapting the FMR connection elements and without them providing enough observations to categorize the advantages of adapting FMR connections in the system.

The composing elements of the test apparatus are:

External framing system, composed of steel H members connected at the ends with hinged connections

Secondary beam elements, connected to the vertical columns of the main framing system again with hinged connections and providing supports for the connections of the curtain wall system

Hydraulic actuator, connected to the top beam of the main frame and providing the desired forces and displacements

The curtain wall mockup, containing all the elements of a full sized spider glazing system and constructed exactly as those used in actual building

FMR connection element, that are placed between the spider brackets of the curtain wall system and the secondary beams of the test chamber

Figure 8.3 demonstrates a schematic shape of the test apparatus and the connections between the curtain wall mockup and the test chamber structure.

Performing the dynamic and statics tests, in two cases—with and without the connection devices, will provide data and information on, the advantages of adapting the FMR connection devices on both safety criteria—preventing mechanical failure in the glass panes—and serviceability criteria—maintaining the air tightness and water tightness functionality of the silicon patches—in-between the connection elements of the spider glazing systems.

8.3 Summary and Conclusions

The goals of this research were set to first define and propose connection systems which result in a compatible mechanical behavior between the building structure and its envelope during earthquake, and second to provide reliable instructions for tuning and adjusting the proposed connection devices, thus insuring that, when adopted within the curtain wall system, they will perform in the expected manner.

Study has been done on different mechanism and approaches that are already used in the design of advanced connectors and other energy dissipating techniques which are incorporated in the structural design of the building, and among these methods friction connection devices have been selected as a proper technical approach to be used for an energy dissipating and isolating connection. This decision has been made based on:

1. Highly predictable behavior
2. Rigid behavior of the connection prior to slippage
3. Ability to control the forces transferred through the connection
4. Simplicity of the manufacturing and installation

Among all the advantages of a friction connection device mechanism, the most important one is to sharply confine the forces that are transferred through the connection device.

Design of a connection device

Among different types of curtain wall glazing systems studied in this research, a great deal of attention was given to structural glazing systems. The reason is the high vulnerability of these systems, and the lack of proper provisions to protect them against, seismic actions. For that a special friction connector device was proposed specifically for these systems to satisfy the main design objectives described below:

Applicability to the implemented in Complex geometrical situations
Satisfy a higher aesthetical demand than the usual advanced connectors in practice.
Ability to control the moments applied on the curtain wall system to avoid out-of-plain deformations

To satisfy the objectives above, instead of using transversal forces caused by displacements in the main structure to trigger frictional behavior, the resulting moments at the two ends of the Connection device are used to carry out this function. Based on this concept the device was named the Friction Moment Rod (FMR). Based on the concepts of applying rotational friction in connection devices, resulting in design of the FMR, another rotational friction device was also designed to be used in other types of curtain wall systems such as stick or unitized systems. This connection device is designed to easily replace the existing isolating connections currently used in these systems.

After having designed the connection device, it is necessary to obtain the mechanical characteristics and adjust the behavior of the connection in a desired way. This was done by investigating the mechanical behavior of different curtain wall systems during earthquake.

Studying the behavior of glass panels during earthquakes

The source of seismic loads that act upon the components of a structure is the acceleration nature of a seismic event which, multiplied by the mass of the structure, results in force between the composing elements. But unlike structural members the source of the forces applied on light envelope systems, due to their comparably low mass, is not the accelerating nature of a seismic event, and the displacement which happens within a structure are the cause of damaging forces. That is why using connection devices that result in compatible behavior between the two systems is of great importance. At this point the mechanical behavior of different curtain wall systems was investigated based on drifts which will happen in a structure. Considering the significantly small dimensions of glass thickness compared to its height and width, and the in-plane nature of the applied forced, the buckling of glass pane was set to be analyzed as the most portable mode of failure in glass components during an earthquake.

The theory of buckling of thin plates subjected to distributed shear force was used for analyzing unitized systems and the result of the analysis implied that; considering the amounts of shear at the edges of the glass plate the buckling failure in these systems will be proceeded by failure in silicon patches and frame distortions. The high values of the critical shear forces in this case were associated to the edge supports of the glass preventing buckling. This type of lateral support is not present in the cases of structural glazing.

For the case of structural glazing systems, due to unavailability of a closed form theoretical solution to the problem at hand, two sets of simplifying assumptions where used for obtaining results. The absence of theoretical solutions that completely match the circumstances of the problem, are first because of existence of point loads in the model, and second due to the free edge boundary conditions of the problem.

The first simplified model was based on an analogy between the system at hand with a plate subjected to shear forces, and the second with a uniaxially loaded plate. Both cases were considered to have free lateral boundary conditions.

It is necessary to indicate that in comparison with the exact solution to the actual problem, both models were suggested in a way to give upper hand solutions to the problem so that the results can be reliable for the design of the connection device. The results of both simplified models have shown to be consistent with each other.

Numerical simulations

In order to verify and evaluate the accuracy of the simplified models produced, a set of numerical analyses has also been performed in this research. At first the two software packages of SAP2000 and SJ-MEPLA were used to perform a static numerical simulation of the problem and a buckling analysis. Since it was not possible to realistically model sandwich shell elements in SAP2000, the recommended corrections of the European standards prEN 13474-3 (European Standards 2009) for thickness of laminated glass components, were used for glass pane inputs to the software. On the other hand being a commercial Finite Element software specifically developed for structural analysis of glass, the SJ-MEPLA had the ability to completely simulate all the details of a structural glazing system including laminated properties and fixing details.

The results of the two numerical simulations have shown very good correspondence for thinner glass elements with one interlayer of PVB (0.38 mm). But for thicker values of glass, with a double interlayer equal to 0.76 mm, the result of SJ-MEPAL for critical buckling forces were between 10 and 25 % less than the ones of SAP2000. This would suggest a further reduction required in the bending stiffness of laminated glass panes with more than one PVB interlayer. Comparing the results of the analytical solutions with the numerical solutions had shown that the results of the analytical solutions can be used as reference loads to determine the limit force value for the design of the Friction Moment Rod with an average safety factor equal to 1.5.

After having determined the limit force values (tuning forces) for the connection device, non-linear time history analysis were performed on a single glass panel connected to supports with friction damping connectors. This type of numerical simulation was possible only with SAP2000 software. In order to demonstrate the effect of lateral drifts on the glass panel a sinusoidal displacement was imposed on the support of the glass panel and link elements with elastic-perfectly-plastic behavior (representing the FMR devices), were located between the glass and the supports. Comparison between the maximum values of the stress fields in the dynamic simulation with near buckling stress states, made clear that adopting friction connection devices with the adjustments based on analytical models, will result in protection of the glass panels against lateral displacement within the structure.

Finally in order to investigate the effects that a set of adjacent structural glazing panels may have on each other, a group of panels connected with silicon elastomer material were modeled in SAP2000 and again subjected to support displacements based on a function uniformly increasing with the elevation of the supports. In this case only minor and somehow negligible increase in the stress state of the lower glass panels was witnessed.

Further suggestions for research

Select materials instead of lining pads more often used in building sector.
Performing numerical simulations in ANSYS or ABAQUS to have both the advantages of SAP and MEPLA at the same time.
Investigating the behavior of silicon patches between glass panes in the dynamic model (for serviceability criterion) from models with a group of panels.
More thorough investigation on the cumulative behavior of the glass panels.

References

Association of American Architectural Manufacturers. (2000). *Recommended static test method for evaluating curtain wall and storefront systems subjected to seismic and wind induced interstory drifts.* Schaumburg: American Architectural Manufacturers Association.

Association of American Architectural Manufacturers. (2001). *Recommended dynamic test method for determining the seismic drift causing glass fallout from a wall system.* Schaumburg: American Architectural Manufacturers Association.

Pinelli, J. P., Moor, C., Craig, J. I., & Goodno, B. J. (1996). Testing of energy dissipating cladding connections. *Earthquake Engineering and Structural Dynamics, 25*(2), 129–147.

Index

© The Author(s) 2015
R. Afghani Khoraskani, *Advanced Connection Systems for Architectural Glazing*,
PoliMI SpringerBriefs, DOI 10.1007/978-3-319-12997-6